Marine Diesel Engines

Maintenance, Troubleshooting, and Repair

Second Edition

Nigel Calder

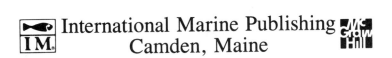

International Marine Publishing
Camden, Maine

To Terrie,
who never minds getting grease under her fingernails

Other Books by Nigel Calder
Repairs at Sea
Refrigeration for Pleasureboats
Boatowner's Mechanical and Electrical Manual
The Cruising Guide to the Northwest Caribbean

Published by International Marine

10 9 8 7 6 5 4 3 2 1

Library of Congress Cataloging-in-Publication Data

Calder, Nigel.
   Marine diesel engines : maintenance, trouble-shooting, and repair /
Nigel Calder.—2nd ed.
     p. cm.
   Includes bibliographical references and index.
   ISBN 0-87742-313-X
   1. Marine diesel motors—Maintenance and repair. I. Title.
VM770.C25 1991   91-31213
623.87'2368—dc20   CIP

Questions regarding the content of this book should be addressed to:

International Marine Publishing
P.O. Box 220
Camden, ME 04843

Typeset by TAB Books, Blue Ridge Summit, PA
Printed by Arcata Graphics, Fairfield, PA
Front cover photo courtesy Caterpillar Tractor Co.

Contents

List of Troubleshooting Charts

An untrained observer will see only physical labor, and often gets the idea that physical labor is what the mechanic does. Actually, the physical labor is the smallest and easiest part of what the mechanic does. By far the greatest part of his [or her] work is careful observation and precise thinking.

Robert M. Pirsig
Zen & the Art of Motorcycle Maintenance

Preface to the Second Edition

It is five years now since I went to work on the first edition of this book. Since then it has been a best-seller, which is very gratifying. In the intervening years, there have been no startling or revolutionary technological breakthroughs in the world of marine diesel engines, but as always there has been a steady and continuing process of improvement and change. This second edition has enabled me to catch up with some of these developments.

Of more significance than changes in the diesel world is the expansion and refinement of my own ideas on a number of the topics covered in this book. The new edition reflects this. In particular, I pay more attention to the electrical side of marine diesel engines—a subject that was given pretty short shrift in the first edition—and I have expanded the section on engine overhauls to include many new illustrations, for most of which I am greatly indebted to William Gardner Ph.D., author and lecturer in the diesel engine field, and the Caterpillar Tractor Co. The coverage of Detroit Diesel two-cycle engines, marine transmissions, and engine installations is also more thorough.

I have extensively rewritten and reorganized the introductory chapters, utilizing material from a series of articles I wrote for *National Fisherman* in 1987—my thanks to them, too. In spite of all these changes, the purpose of the book is unchanged: to provide the basic information necessary to select, install, maintain, and carry out repairs on a marine diesel engine. As such, the book remains neither a simple how-to book nor a technical manual on the thermodynamics of internal-combustion engines. Rather, it falls somewhere between the two.

As I stated in the first edition, this reflects my own experience as a self-taught mechanic with some 25 years experience on a variety of engines from 10 to 2,000 h.p. With specific enough instructions it is possible to dismantle an engine and put it back together again without having any understanding of how it works, but troubleshooting that engine is not possible without a basic grasp of its operating principles.

In order to grasp these operating principles, you need to know a little of the most basic theory behind internal-combustion engines. This I try to present, but only insofar as it is necessary to provide a good understanding of the practical side of diesel engine maintenance. My objective is to help turn out competent amateur mechanics, not automotive engineers.

This particular blend of theory and practical mechanics represents a mix that has worked well for me over the years. It should help a boatowner to see trouble coming and to nip it in the bud before an engine breaks down.

Although this book has been written with engines from 10 to 100 h.p. in mind, the principles are virtually the same as those associated with engines of hundreds or even thousands of horsepower. The information in this book applies to just about all diesels.

Sources of data and drawings are indicated throughout the book, but nevertheless my thanks to all those who have helped me, in particular Paul Landry,

Bill Osterholt, and Dennis Caprio, my editor, who made a mass of detailed suggestions that improved the book greatly. The following companies provided drawings and other help: the AC Spark Plug Division of General Motors, Allcraft Corporation, Aquadrive, Borg Warner Automotive, Caterpillar Tractor Co., Deep Sea Seals, Detroit Diesel, Garrett Automotive Products Co., Halyard Marine, Hurth, Hart Systems Inc., Holset Engineering Co. Ltd., ITT/Jabsco, Kohler Generators, Lucas CAV Ltd., Morse Controls, Paragon Gears, PCM, Perkins Engines Ltd., PYI, Racor, Sabb Motor A.S., Shaft Lok, United Technologies Diesel Systems, VDO, Volvo Penta, Wilcox Crittenden, and Yanmar Diesel Engines. Many of the line drawings were done by Jim Sollers, a wonderful illustrator.

For the second edition, Caterpillar Tractor Co. once again dug deep into its incomparable educational material. Don Allen of Allen Yacht Services (St. Thomas, USVI), Russell Dickinson, Joseph Joyce of Westerbeke, and Steve Cantrell of Volvo Penta all critiqued the text and made many valuable comments.

I extend my thanks to Dodd, Mead & Company for permission to use the material from Francis S. Kinney's *Skene's Elements of Yacht Design*, and to Reston Publishing for information contained in Robert N. Brady's *Diesel Fuel Systems*. My publisher, International Marine, gave me permission to incorporate one or two relevant sections from my *Boatowner's Mechanical and Electrical Manual* in this book, and information from Dave Gerr's *Propeller Handbook* (quite the best book available on this subject). Sections of the revised chapter on "Engine Installations" first appeared in an article I wrote for *WoodenBoat* magazine.

Jonathan Eaton, Tom McCarthy, Janet Robbins, and the crew at International Marine have been as helpful and encouraging as ever. Any errors remaining are solely mine.

Nigel Calder,
Montana,
November 1990

Introduction

For very good reasons the diesel engine is now the overwhelming choice for sailboat auxiliaries, and it is becoming more popular in sportfishing boats, high-performance cruisers, and large sport boats. Diesels have an unrivaled record of reliability in the marine environment; they have better fuel economy than gasoline engines; they are more efficient at light and full loads; they emit fewer harmful exhaust pollutants; they last longer; and they are inherently safer because diesel fuel is less volatile than gasoline.

Despite its increasing popularity, the diesel engine is still something of a mystery, in large part because of the differences that distinguish it from the gasoline engine. Although many boatowners are quick to tackle quite complicated problems on their automobiles, they are frequently nervous about carrying out maintenance and repairs on a diesel. This is surprising because in many respects diesels are easier to understand and maintain than gasoline engines. The first objective of this book, then, is to explain how a diesel engine works, to define new terms, and to lift this veil of mystery.

If the owner of a diesel engine has a thorough understanding of how it works, the necessity for certain crucial aspects of routine maintenance and the expensive consequences of habitual neglect will be fully appreciated. Properly maintained, most diesel engines will run for years without trouble, which leads me to my second objective—to drive home the key areas of routine maintenance.

If and when a problem arises, it normally falls into one or two easily identifiable categories. The ability to visualize what is actually going on inside an engine frequently enables you to go straight to the heart of a problem and to rapidly find a solution, without making blind stabs at it. My third objective is to outline troubleshooting techniques that promote a logical, clearheaded approach to problem solving.

The fourth section of the book goes into various maintenance, overhaul, and repair procedures that can be reasonably undertaken by an amateur mechanic. It also points out one or two that should not be attempted but which might become necessary in a dire emergency. (Cruising sailors, in particular, sometimes must tackle repairs that they would never even consider when they are ashore. As a fellow sailor, I have at all times tried to keep the rather special needs of cruisers in the forefront of my mind when writing and revising this book.) Major mechanical breakdowns and overhauls are not included. This kind of work can only be carried out by a trained mechanic.

The book is rounded out with a consideration of some criteria to assist in the selection of a new engine for any given boat, and of correct installation procedures. Much of the last section may throw light on problems with an engine already installed.

There is no reason for a boatowner not to have a long and troublefree relationship with a diesel engine. All one needs is to set the engine up correctly in the first place, to pay attention to routine maintenance, to have the knowledge to spot early warning signs of impending trouble, and to have the ability to correct small problems before they become large ones.

Chapter 1

Principles of Operation

In the technical literature, diesel power plants are known as compression ignition (CI) engines. Their gasoline counterparts are of the spark ignition (SI) variety. This idea of *compression ignition* is central to understanding diesel engines.

When a given amount of any gas is compressed into a smaller volume, its pressure and temperature rise. The increase in temperature is in direct relation to the rise in pressure, which is directly related to the degree of compression. The rise in temperature is not due to the addition of any extra heat but is simply the result of confining the existing heat of the gas in a smaller space.

For a better understanding, imagine two heaters with exactly the same output. Each has been placed in a separate room. To begin with, both rooms are the exact same temperature, but one is twice the size of the other. Both heaters are turned on. The small room will heat up faster than the larger one, even though the output of the heaters is the same. In other words, although the same quantity of heat is being added to both rooms, the temperature of the smaller one rises faster because the heat is concentrated into a smaller space.

This is crudely analogous to what happens when a gas is compressed. At the outset, it has a given volume and contains a certain amount of heat. As the gas is compressed, this quantity of heat is squeezed into a smaller space and the temperature rises, even though no more heat is being added to the gas.

Compression Ignition

All internal-combustion engines consist of one or more cylinders that are closed off at one end and have a piston driving up the other. In a diesel engine, air enters the cylinder, then the piston is forced up it, compressing the air.

As the air is compressed, the heat contained in it is squeezed into a smaller and smaller space, which causes the pressure and temperature to rise steadily. In a compression ignition engine, this procedure continues until the air is extremely hot, say around 1,000 °F (538 °C). This temperature has been attained purely and simply by compression (see Figure 1-1).

Diesel fuel ignites at around 750 °F (399 °C); therefore, any fuel sprayed into a cylinder filled with 1,000 °F super-heated air is going to catch fire. This is exactly what happens: at a precisely controlled moment, fuel is *injected* into the cylinder and immediately starts to burn. No other form of ignition is needed.

In order to attain high enough temperatures to produce ignition of the diesel fuel, air generally has to be compressed into a space no larger than $1/14$ the original size of the cylinder. This is known as a *compression ratio* of 14:1. The compression ratio is the volume of the cylinder when the piston is at the bottom of its stroke relative to the volume of the cylinder when the piston is at the top of its stroke (see Figure 1-2).

1

Most diesel engines have compression ratios ranging from 16:1 to 23:1. This is much higher than the average gasoline engine's compression ratio of 7:1 to 9:1. The lower compression ratios of gasoline engines produce lower cylinder pressures and temperatures. As a consequence, the ignition temperature of gasoline is not reached through compression, and the gas/air mixture has to be ignited by an independent source—a spark (hence the designation spark ignition for gasoline engines).

Converting Heat to Power

I've established that when a gas is compressed its temperature rises. It is also true that when a gas is heated in a sealed chamber its pressure rises. In the first instance, no heat is added—the existing heat of the gas is merely squeezed into a smaller space, thereby raising its temperature. In the second instance, heat is actually being added to a gas that is trapped in a closed vessel, and this causes the pressure to rise.

This is what happens during ignition in an internal-combustion engine: a body of air is trapped in a cylinder by a piston and compressed. The temperature rises. Fuel is introduced by some means and ignited. The burning fuel raises the temperature in the cylinder even higher, and this raises the pressure of the trapped gases. The increased pressure is used to drive the piston back down the cylinder, resulting in what is termed the piston's *power stroke*. The engine has converted the heat produced by the burning fuel into usable mechanical power. For this reason, internal-combustion power plants are sometimes known as *heat engines*.

It is possible to calculate the heat content of the fuel by measuring how many Btus (the unit of measurement of heat) are produced by burning one gallon. An engine's horsepower can also be converted into Btus—one h.p. equals 2,544 Btus. In this way, the heat energy going into an engine can be compared with the power being produced by it. This enables the *thermal efficiency* of the engine to be determined—how much of the fuel's heat energy is being converted to *usable* power.

The average diesel engine has a thermal efficiency of 30% to 40%. In other words, only about one-third of the heat energy contained in the fuel is being converted to usable power. Roughly half of the remaining

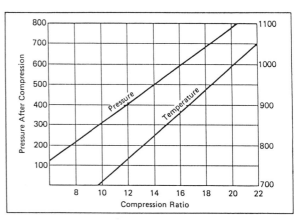

Figure 1-1. *Approximate temperatures and pressures at different compression ratios.*

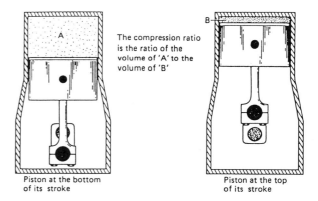

Figure 1-2. *Compression ratios. (Courtesy National Fisherman)*

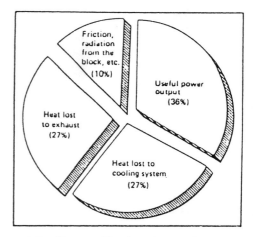

Figure 1-3. *Heat utilization in a diesel engine. (Courtesy National Fisherman)*

two-thirds is lost through the exhaust system in the form of hot gases. The other half is dissipated into the atmosphere through the cooling system and by contact with hot engine surfaces (see Figure 1-3). As wasteful as this sounds, diesels are still considerably more efficient than gasoline engines, which have a thermal efficiency of 25% to 35%.

Expansion and Cooling

Just as compressing a gas raises its temperature, so too can reducing the pressure lower the temperature. This is due purely to the expansion of the gas into a larger space, not to any loss of heat. The greater the reduction in pressure, the lower the resulting temperature of the gas.

As a piston moves down on its power stroke, the volume inside its cylinder increases, causing a fall in pressure and consequently a fall in the temperature of the gases in the cylinder. These declining temperatures reflect the heat of combustion being converted into mechanical power; i.e., the movement of the piston (see Figure 1-4).

The higher the compression ratio of an engine, the greater the expansion of gases on the power stroke. In an engine with a compression ratio of 22:1, for example, the gases will expand into a volume 22 times the size of the compression chamber. In an engine with a compression ratio of 7:1, the degree of expansion will only be seven times greater.

Since the temperature drops as a gas expands, diesel engines, because of their higher compression ratios, are able to convert more of the heat of combustion into mechanical power than their gasoline counterparts. Hence diesels are more thermally efficient than gasoline engines.

Gasoline Engines

You might well ask: Why not increase the compression ratio on the gasoline engine and thereby improve its efficiency?

A gasoline engine draws in fuel with its air supply *before* compression, either through a carburetor or through fuel injection *into the inlet manifold* (not the cylinder). A diesel, on the other hand, has the fuel

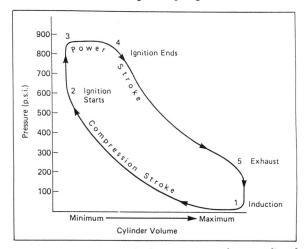

Figure 1-4. *Pressure/volume curve for a diesel engine. From position 1 (P1) to position 2 (P2) the pressure steadily rises to around 600 psi as cylinder volume decreases. At around P2, injection occurs and ignition commences. The rapidly rising temperature sharply boosts pressure to about 850 psi (P3). The piston is now on its way back down the cylinder in its power stroke, increasing the volume and decreasing pressure. However, fuel is still burning, and the increasing temperature temporarily counteracts the effect of increasing cylinder volume, which causes a relatively constant pressure from P3 to P4. Ignition is now tailing off, and the cylinder volume continues to increase. This results in a steady decline in pressure and temperature while the piston is still on its power stroke from P4 to P5. From P5 to P1 the engine expels the remaining gases of combustion as exhaust and draws in a fresh charge of air. The cycle starts over.*

injected *after* the air has been compressed. Increasing the compression ratio for a gasoline engine would raise its compression temperature beyond the ignition point of gasoline, which would lead to premature combustion of the fuel/air mixture. This would rapidly wreck the engine. To avoid this, the compression ratio on a gasoline engine must be kept low and the fuel/air mixture must then be set off by a spark at the appropriate moment. This accounts for the need for an electrical ignition system on these engines.

On occasion, gasoline engines can become sufficiently overheated to cause the fuel/air mixture to ignite before it should. This is known as *auto-igni-tion*, or *pre-ignition*, and most often occurs when the overheated engine is turned off but refuses to quit, even though the ignition has been turned off.

You might next ask: to improve efficiency why not jack up the compression ratios on gasoline engines and use fuel injection directly into the cylinders to prevent premature ignition, just as is done on a diesel? Apart from the obvious fact that you now have a diesel engine to all intents and purposes and might as well use the cheaper diesel fuel, you run up against the nature of gasoline itself. This fuel is far more volatile than diesel fuel.

Even though a diesel engine may be turning over at 3,000 r.p.m. (revolutions per minute), with the power stroke of any one piston lasting no more than $1/100$ of a second, the injected diesel fuel burns at a *controlled rate*, rather than exploding. Indeed, if it fails to burn at the correct rate, ignition problems result and engine damage is likely.

Because of its greater volatility, the same degree of control cannot be maintained over gasoline. Explosive combustion would occur, destroying the high-compression power plant. The gasoline engine, at the

current level of technology, is therefore locked into lower compression ratios and decreased thermal efficiency.

Cost and Power-to-Weight

Because diesels feature higher compression ratios than gasoline engines, they are subjected to greater stresses and must be built more ruggedly. To sustain these higher compression ratios and loads, diesels generally have to be machined to closer tolerances. The heavier construction and closer machining tolerances account for the increase in weight and price of diesel engines over gasoline engines of the same power output. In recent years, however, tremendous advances in metallurgy and engine design have enabled drastic weight reductions to be achieved on many diesels, considerably narrowing this power-to-weight gap.

Types of Diesels

Diesel engines can be *4-cycle* or *2-cycle*. The differences will become clear in a moment. Let us first look at a 4-cycle engine.

A 4-Cycle Diesel

1. Imagine the piston is at the top of its cylinder. The inlet valve opens as the piston descends to the bottom of its cylinder. The descending piston draws air into the cylinder. When the piston reaches the bottom of its cylinder, the inlet valve closes, trapping the air inside the cylinder (see Figure 1-5). This movement of the piston from the top to the bottom of its cylinder is known as a *stroke*. It also constitutes one of the four cycles of a 4-cycle engine—in this case the suction (*induction*), or inlet, cycle.

2. The piston now travels up the cylinder, compressing the trapped air. The pressure rises to between 450 psi and 700 psi (as compared to 80-150 psi in a gasoline engine) and the temperature to 1,000 °F (538 °C) or more. This is the compression cycle.

3. Somewhere near the top of the compression stroke, fuel enters the cylinder via the fuel injector and starts to burn. The temperature

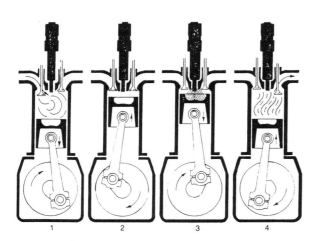

Figure 1-5. *The four cycles of a 4-stroke diesel engine. (1) Inlet stroke. The piston draws air into the cylinder via the inlet valve. (2) Compression. The piston compresses the air, which also heats it. (3) Injection. Fuel sprayed into the hot air ignites and burns in a controlled manner. (4) Exhaust. The piston forces burnt gases out the exhaust-valve opening. (Courtesy Lucas CAV Ltd.)*

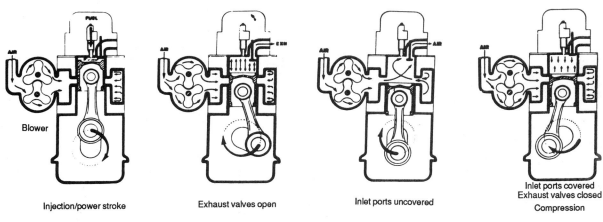

Figure 1-6. *The operation of a 2-cycle Detroit Diesel. (Courtesy Detroit Diesel Corp.)*

climbs rapidly to anywhere from 2,000 °F (1093 °C) to 5,000 °F (2760 °C). This increase in temperature causes a rise in pressure to around 850 to 1,000 psi, pushing the piston back down its cylinder. As the piston descends, the cylinder volume increases rapidly, leading to a sharp reduction in the pressure and temperature. This is the third cycle and is known as the power stroke.

4. When the piston nears the bottom of the power stroke, the exhaust valve opens. The cylinder still contains a considerable amount of residual heat and pressure, and most of the gases rush out. The piston then travels back up the cylinder, forcing the rest of the burned gases out of the exhaust valve. This is the fourth, or exhaust cycle.

At the top of the exhaust stroke, the exhaust valve closes and the inlet valve opens, ready to admit a fresh charge of air when the piston descends the cylinder once again. This brings the engine back to the starting point of the four cycles.

A 2-Cycle Diesel. Note: the following is a description of the operation of a Detroit Diesel 2-cycle. These are the most common and widely known. There are engines with other forms of 2-cycle operation, but such engines are unlikely to be encountered in small-boat applications.

A Detroit Diesel 2-cycle engine operates in essentially the same manner as a 4-cycle engine, but condenses the four strokes of the piston into two—once up the cyclinder and once down. Here's how:

1. We start with the piston at the top of its cylinder on its compression stroke. The cylinder is filled with pressurized, superheated air. Diesel is injected and ignites. The piston starts back down the cylinder on its power stroke. As it descends, the cylinder pressure and temperature fall. When the piston nears the bottom of its power stroke, the exhaust valve opens and most of the burned gases rush out of the cylinder (see Figure 1-6). So far all is the same as for a 4-cycle diesel.

Now as the piston continues to descend the cylinder, it uncovers a series of holes, or *ports*, in the cylinder wall. A supercharger or turbocharger blows pressurized air through these ports, pushing the rest of the burned gases out of the cylinder and refilling it with a fresh air charge. The piston has only now reached the bottom of its cylinder and is starting back up again. The exhaust valve closes.

2. As the piston moves back up, it blocks off the inlet ports, trapping the charge of fresh air in the cylinder. Although the piston has only covered a little over one stroke, it has already completed its power stroke, the exhaust process, and the inlet cycle. As the piston comes back up the cylinder on its second stroke, it compresses the fresh air. When it reaches the top of the cylinder, injection and combustion take place. The cycle starts over. The engine

has done in two strokes what a 4-cycle diesel does in four.

A 2-cycle engine, therefore, has two power strokes for every one of a 4-cycle engine. For a given engine size, a 2-cycle engine develops considerably more power than a 4-cycle. This leads to lower costs per horsepower and improved power-to-weight ratios.

A 2-cycle diesel, however, is less thermally efficient than a 4-cycle, and fuel consumption is higher. The life of a 2-cycle diesel tends to be shorter than that of a 4-cycle model because of the higher loads placed on the engine. What's more, for reasons that will become clear later on, 2-cycle diesels tend to be far noisier in operation than 4-cycles, which makes them unsuitable for a wide range of pleasureboat applications.

The Crankshaft

So far I have talked in a purely abstract fashion of a piston moving up and down its cylinder. To harness a piston to the rest of the engine, and to utilize the mechanical energy developed by its power stroke, this reciprocal motion must be converted to rotary motion. This is done by a *crankshaft* and *connecting rod*.

A crankshaft is a sturdy bar set in bearings in the base of an engine. Beneath each cylinder, it has an offset pin, forming a crank. The connecting rod ties the piston to this crank. A bearing at each end of the connecting rod allows the crank to rotate within the connecting rod's lower end, while the piston, mounted on a *piston pin* or *wrist pin*, oscillates around its upper end. As the piston moves up and down, the crankshaft turns (see Figure 1-7).

Valves and Timing

The effective operation of 4-cycle and 2-cycle engines requires the precise coordination of piston movement with valve opening and closing times, as well as with the moment of fuel injection. This is known as valve and fuel-injection timing.

Valves are set in cylinder heads and held in a closed position by a valve spring. A lever known as a *rocker arm* opens the valve. The rocker arm, moved up and down directly or indirectly by a *camshaft*, pivots in the top of the cylinder head (see Figure 1-8).

A camshaft has along its length a series of elliptical protrusions, or *cams*, (one for each valve in the engine). As the camshaft turns, these protrusions push the rocker arms up and down. Some camshafts are set in cylinder heads with the cams in direct contact with the rocker arms—these are *overhead camshafts*. Others are placed in the engine block and actuate the rocker arms indirectly via *push rods*.

A gear, keyed (locked) to the end of the crankshaft, rotates with it. Another gear, keyed to the end of the camshaft, rotates with it. The valve timing, or valve opening and closing times, is coordinated with the movement of the pistons by linking the two gears

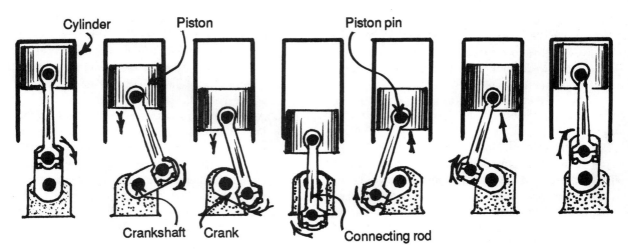

Figure 1-7. *Converting reciprocal motion to rotary motion.*

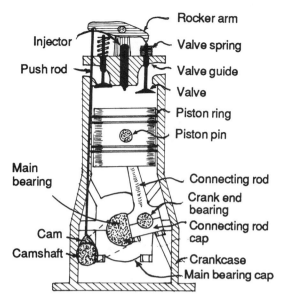

Figure 1-8. *Principal components in a diesel engine.*

Figure 1-9. *Engine timing gears. (Courtesy Caterpillar Tractor Co.)*

with an intermediate gear, a belt, or a chain so that they rotate together (see Figure 1-9).

On a 4-cycle engine, the inlet and exhaust valves are opened and closed on every other stroke of the engine. As a consequence, the gear on the camshaft is twice the size of that on the crankshaft, causing the former to rotate at half the speed of the latter and the valves to open and close on every other engine revolution.

Detroit Diesel 2-cycle engines have only exhaust valves. The exhaust valves open on every downward stroke of the piston; therefore, the camshaft gear is the same size as that on the crankshaft. The two shafts rotate at the same speed, and the valves open and close on every engine revolution.

By setting the camshaft and crankshaft gears in different relationships with one another, the valves can be made to open and close at any point of piston travel. Timing engine valves consists of placing the gears in the precise relationship needed for overall maximum engine performance.

Fuel injection timing is set the same way. A gear fitted to the end of the fuel-injection pump drive shaft is driven by the crankshaft via an intermediate gear, belt, or chain. Altering the relationship of the gears allows the fuel injection to be set at any point of piston travel. Since injection occurs on every other revolution on a 4-cycle engine, the pump drive gear is twice

the size of the crankshaft gear, causing the pump to rotate at half the engine's speed. On 2-cycle engines, the gears are the same size, producing injection at every engine revolution.

Cylinders and Other Parts

The central frame to which the rest of an engine is assembled is known as the *block*, or *crankcase* (see Figure 1-10). On all except air-cooled engines, which are rarely found on boats, this is a casting containing numerous passages for air, cooling water, oil, and other engine parts, such as the crankshaft and camshaft.

There are two types of cylinders, or *liners*, available for diesel engines—dry liners and wet liners. In a dry-liner diesel the cylinders and block are in a continuous metal-to-metal contact. In a wet-liner arrangement the block comes into contact with the cylinders only at their tops and bottoms, and engine cooling water circulates directly around the liners (see Figure 1-11). Wet liners have the advantage of being relatively easy to replace in the field when a major engine overhaul is necessary, whereas the whole block must generally be taken to a machine shop to change dry liners.

Pistons are fitted with a number of *piston rings*, set in grooves cut into the outside circumference of the

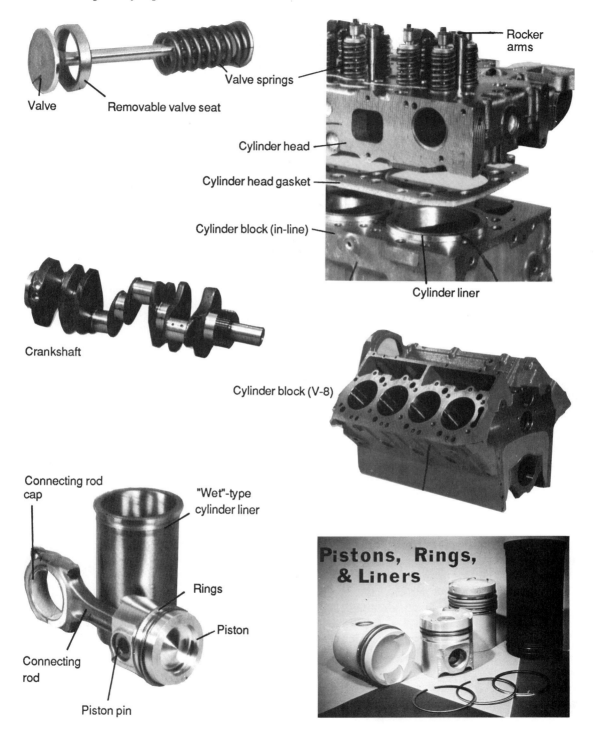

Figure 1-10. *Engine parts. (Courtesy Caterpillar Tractor Co.)*

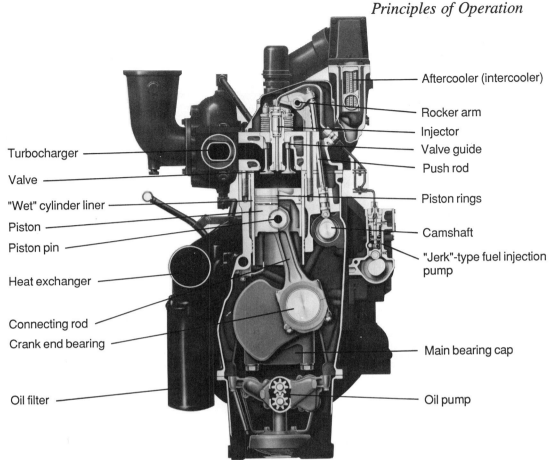

Aftercooler (intercooler)
Rocker arm
Injector
Valve guide
Push rod
Piston rings
Camshaft
"Jerk"-type fuel injection pump
Main bearing cap
Oil pump

Turbocharger
Valve
"Wet" cylinder liner
Piston
Piston pin
Heat exchanger
Connecting rod
Crank end bearing
Oil filter

Figure 1-12. *Cutaway view of a Caterpillar 3406B turbocharged in-line 6 represents typical modern diesels. (Courtesy Caterpillar Tractor Co.)*

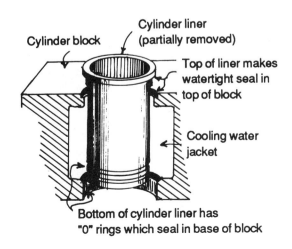

Cylinder block
Cylinder liner (partially removed)
Top of liner makes watertight seal in top of block
Cooling water jacket
Bottom of cylinder liner has "0" rings which seal in base of block

Figure 1-11. *A "wet" cylinder liner.*

piston. The piston rings press out against the walls of the cylinder to make a gas-tight seal. The top end of a cylinder is sealed with a casting known as a *cylinder head*, which contains the injectors, valves, and frequently combustion chambers, water passages, and so on. The valves are set in *guides*, which are replaceable sleeves pressed into the cylinder head.

When an engine begins to show appreciable wear, these valve guides are easily pressed out and replaced. Better-quality engines also have replaceable valve *seats*, the area that a valve contacts to seal the combustion chamber. Re-machining valve seats, replacing valve guides, and installing new valves renews the cylinder head, giving this expensive part an almost indefinite life.

Chapter 2

Details of Operation

SECTION ONE: THE AIR SUPPLY

A diesel engine develops power by burning fuel. The more fuel that an engine of any given size can burn, the more heat it will generate, and the greater will be its power output. Increasing the power-to-weight ratio reduces the cost per horsepower produced. Manufacturers are therefore continually striving to increase the amount of fuel a given engine can burn.

Getting more diesel into a cylinder is no problem—all it takes is a bigger fuel injection pump and injector. Getting it to burn is another matter. Incomplete combustion lowers fuel economy and causes excessive exhaust pollutants. For effective combustion, three interdependent factors must be present:

1. an adequate supply of oxygen;
2. effective *atomization* of the injected diesel;
3. thorough mixing of the atomized diesel with the oxygen in the cylinder.

Let us look first at the oxygen supply (atomization of the fuel and proper mixing with the oxygen are covered in the next section on combustion, which follows).

The Oxygen Supply

What actually takes place when diesel fuel burns is a reaction between oxygen in the air and hydrogen and carbon in the diesel fuel. Starting this reaction takes a temperature of around 750°F (399°C). Once it begins, the oxygen combines with the hydrogen to form water, and combines with the carbon to form carbon dioxide (and occasionally carbon monoxide when combustion is incomplete). In the process of these chemical reactions, considerable quantities of heat and light are released—the fuel *burns*.

Air is only about 23% oxygen *by weight* (21% by volume); the rest is principally nitrogen, plus one or two trace gases, and plays no part in the combustion process. This idea of air having weight may be a little confusing, so let's take a moment to look at this.

You are probably familiar with the concept of *atmospheric pressure*, the notion that the weight of the atmosphere exerts pressure on the surface of the earth. As you climb to higher elevations, the atmosphere becomes thinner, i.e., it exerts less pressure. At the top of Mount Everest, for example, you need an oxygen mask to breathe. Deep space has no atmosphere—it is a vacuum, or has no pressure whatsoever.

At sea level, 1 cubic foot of air weighs approximately 0.076 pounds at 60°F (15.6°C). At greater altitudes it weighs less. It also weighs less at higher temperatures, because as the temperature rises, air expands so that each cubic foot contains less of it.

To completely burn 1 pound of diesel fuel requires approximately $3^{1}/_{3}$ pounds of oxygen. Because air is only 23% oxygen, burning 1 gallon of diesel fuel at sea level requires almost 1,500 cubic feet of 60°F air. This is as much air as is contained in a good-size room! At higher elevations and temperatures it can take quite a lot more. The ability to get sufficient air into an engine for complete combustion of the fuel becomes the limiting factor in how much fuel the engine can burn.

Volumetric Efficiency

Engineers must do everything possible to avoid restricting the flow of air to an engine. Air filters are made as large as possible, and manufacturers recommend changing them frequently. Inlet manifolds are made as smooth and direct as possible in order to reduce friction with the incoming air. Valves are made as large as can be accommodated in the cylinder head (on 2-cycle diesels the inlet ports are given as large an area as possible). The unavoidable inefficiencies created by the remaining friction in the air passages (including the exhaust—more on this later) are referred to as *pumping losses*.

The degree to which an engine succeeds in completely refilling its cylinders with fresh air is known as its *volumetric efficiency*. From the bottom of its stroke to the top, a piston occupies, or displaces, a certain volume. This is known as its *swept volume*. If an engine were able to draw in enough air on its inlet stroke to completely fill this swept volume *at atmospheric pressure*, it would have a volumetric efficiency of 100%. Volumetric efficiency, then, is the proportion of the volume of air drawn in, relative to the swept volume, at atmospheric pressure.

Naturally Aspirated Engines

An engine that draws in its air charge through the action of its pistons is known as *naturally aspirated*. Here's how it works.

On a standard 4-cycle engine, the downward movement of the piston on the inlet cycle reduces the pressure in the cylinder and pulls air into the cylinder. (Strictly speaking, reduced pressure in the cylinder causes the higher atmospheric pressure on the outside to *push* air into the engine, but it is easier to visualize it as the piston sucking in air.)

Friction caused by the air filter and air-inlet piping (manifold) partially obstructs the flow of air into the engine. As a consequence, when the piston reaches

the bottom of its stroke, the pressure inside the cylinder is still marginally below that of the atmosphere— in other words, there is a slight vacuum. This means that the cylinder has not been completely refilled with air under atmospheric pressure.

As the air rushes into a cylinder, however, it gains momentum. When a piston of a 4-cycle engine reaches the bottom of its stroke on the inlet cycle and starts back up on the compression stroke, air continues to enter the cylinder but only for a moment. In order to take advantage of this, the inlet valve is set to close a little after the piston has started its compression stroke.

The same valve is set to open a short time before the piston has reached the top of its stroke on the exhaust cycle, before the exhaust valve is fully closed. This is known as *valve overlap*, and the two valves are *rocking* at this point. This ensures that the inlet valve

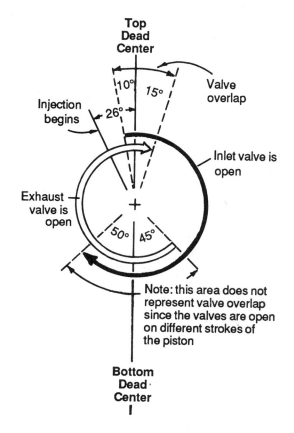

Figure 2-1. *Typical timing circle for a 4-cycle engine.*

is wide open by the time the piston begins its induction stroke, which enables the piston immediately to start drawing in air. These techniques can usually bring up volumetric efficiency to 80% to 90% (see Figure 2-1).

Superchargers and Turbochargers

The volumetric efficiency, and therefore the power output, of a naturally aspirated diesel can be increased dramatically by forcing more air into it under pressure. This is the principle of *supercharging* and *turbocharging*.

A supercharger pumps air into the inlet manifold by means of a fan, or blower, that is *mechanically* driven off the engine via a belt, chain, or gear. A turbocharger consists of a *turbine* installed in the exhaust manifold of an engine and connected to a *compressor wheel* in the inlet manifold. As the exhaust gases rush out, they spin the turbine. The turbine spins the compressor wheel, which pumps air into the inlet manifold. A turbocharger has no mechanical drive (see Figure 2-2).

When the load increases on an engine, more fuel is injected, leading to a rise in the volume of the exhaust gases. This spins the turbocharger's exhaust turbine and compressor wheel faster, forcing more air into the engine. A turbocharger is extremely responsive to changes in load, driving up the power output of an engine just when it is needed most.

A turbocharger won't work on a 2-cycle diesel, at least not unless it's used in conjunction with a supercharger, because when the inlet ports are uncovered by the descending pistons, a 2-cycle diesel depends on a supply of pressurized air to blow the exhaust gases out of its cylinders and refill them with fresh air. This process is called *scavenging*. Since during start-up the engine doesn't have any exhaust gases to spin a turbocharger, it needs a *mechanically driven* blower (a supercharger) to pump air into the cylinders and get everything moving (see Figure 2-3).

The efficiency with which fresh air is introduced to the cylinders in a 2-cycle diesel is known as *scavenging efficiency*, which is much the same concept as volumetric efficiency on 4-cycle engines. If the inlet air drives all the exhaust gases out of a cylinder and completely refills it with fresh air at atmospheric pressure, the engine has 100% scavenging efficiency.

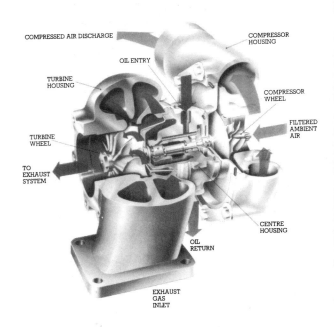

Figure 2-2. *Cutaway view of a turbocharger. (Courtesy Garrett Automotive Products Co.)*

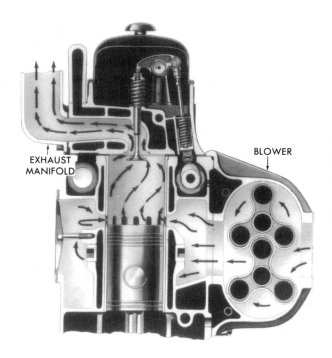

Figure 2-3. *"Scavenging" on a 2-stroke diesel. (Courtesy Detroit Diesel Corp.)*

Intercoolers and Aftercoolers

Air compressed by a turbocharger or supercharger heats up. Because hot air at a given pressure weighs less than cold, it contains less oxygen per cubic foot. In order to counteract the turbocharger's and supercharger's loss of efficiency caused by this rise in temperature, the air must be cooled. Most supercharged or turbocharged engines accomplish this with a heat exchanger, known as an *intercooler* or *aftercooler*, fitted in the inlet manifold between the supercharger or turbocharger and the engine block. (An *intercooler* is fitted between a turbocharger and a supercharger; an *aftercooler* is fitted between a supercharger or turbocharger and the engine block. The two terms tend to get used interchangeably, so in order to avoid cumbersome dual references, I will use aftercooler.) Cooling water is circulated through these units, much like the radiator on your automobile, lowering the temperature of the air as it passes into the engine.

Some aftercoolers are plumbed into the engine cooling circuit, receiving water that has already been warmed by circulation through the engine. Others are plumbed directly to a separate *raw water* supply (see the section on engine cooling). This latter arrangement produces the maximum possible drop in the temperature of the inlet air, therefore allows the greatest possible amount of air to be pumped into the cylinders. Manufacturers will sometimes list three separate horsepower ratings for the same engine to reflect these differing arrangements: 1) naturally aspirated; 2) turbocharged and aftercooled using the engine water circuit; and 3) turbocharged and aftercooled using a separate water circuit.

Superchargers and turbochargers frequently will raise volumetric efficiency to 150%, or more, which is to say that the air in the cylinder at the end of the induction stroke is well above atmospheric pressure. Such an engine commonly develops 50% more power than the same-sized naturally aspirated diesel.

Of course, the price paid for this improved efficiency is more complex and expensive engines (although cost per horsepower is frequently less). Also, turbocharging and supercharging accelerate engine wear and increase the costs of servicing. But the considerable improvement in power-to-weight ratios is a major benefit to many weight-conscious boatowners.

SECTION TWO: COMBUSTION

When diesel fuel is injected into a cylinder containing high-pressure superheated air, it does not explode—it burns. The relatively slow burning of diesel fuel produces a more even rise in cylinder temperatures and pressures than does gasoline, exerting a more gradual force on the piston over the whole length of its power stroke. This is one of its advantages over gasoline, and as a result, diesel engines have far more constant *torque* (the turning force exerted by the crankshaft), especially at low speeds.

The Importance of Turbulence

At the moment of injection, pressure in a cylinder can be as high as 700 psi. The temperature can be more than 1,000°F (538°C). An *injector* sprays into this dense mass of superheated air fuel in the form of one or more streams of minute particles. Only a little more than 20% of the air is oxygen. When a particle of diesel encounters an oxygen molecule, it begins to burn, consuming the oxygen in the process. Complete combustion requires the fuel to come into contact with fresh oxygen, but as the burning progresses, fewer and fewer oxygen molecules remain in the cylinder. Those that do remain are not always in the right place.

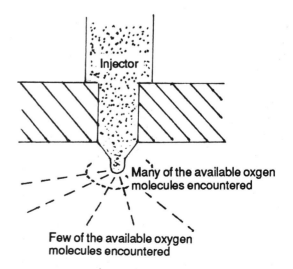

Figure 2-4. *Oxygen utilization.*

Failure to engage a molecule of oxygen causes the half-burned diesel to be blown out of the exhaust pipe as black smoke, lowering fuel economy and robbing power from the engine.

Injectors can spray fuel only in straight lines. When the particles of diesel first emerge from the tip of an injector, they are fairly densely packed and encounter most of the oxygen molecules. As they move farther from the injector, they spread out, and fewer of the oxygen molecules lie directly in their path (see Figure 2-4). Because of this, a mixing together of the fuel and air is imperative.

This business of mixing fuel and air is highly scientific. It has been the subject of much investigation and experimentation over the years, and in modern engines several approaches are taken. The principal variations lie in the nature of the pattern in which the diesel fuel is injected into the cylinders, and in the techniques used for creating turbulence inside the cylinders so that the air and diesel become thoroughly mixed.

Injector Spray Patterns

The fuel injection spray pattern is determined by the size and shape of the openings in the injector *nozzle*, of which there are two basic styles.

- *Hole-type nozzles* force the fuel through one or more tiny orifices. Varying the size of the hole(s) atomizes (breaks up) the fuel to a greater or lesser extent. Changing the number and angles of the holes projects the fuel to various parts of the combustion chamber (see Figure 2-5).
- *Pintle (pin) nozzles* inject a conical pattern of fuel out of a central hole down the sides of a pin (pintle). This type cannot atomize the fuel to the same degree as a hole-type nozzle. Varying the angle of the side of the pintle makes the cone of fuel larger or smaller.

There are also various hybrid injectors, such as the *Lucas CAV Pintaux*, which consists of a pintle nozzle with an auxiliary hole.

Pintle nozzles have a major advantage over hole nozzles: the action of the fuel scrubbing down the sides of the pintle helps to keep it clean, whereas the tiny orifices in hole-type nozzles can be blocked by even the smallest piece of debris.

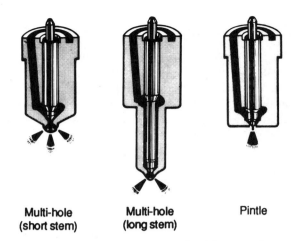

Multi-hole (short stem)   **Multi-hole (long stem)**   **Pintle**

Figure 2-5. *Types of injector nozzle. (Courtesy Lucas CAV Ltd.)*

Techniques for Creating Turbulence

In order to accomplish the mixing of the injected fuel particles with the oxygen in the cylinders, manufacturers design pistons and combustion chambers in ways that impart a high degree of turbulence to the air in the cylinders. Almost all modern diesels use one of the following designs.

Direct Combustion Chambers. A direct, or open, combustion chamber is really no more than a space left at the top of a cylinder when the piston is at the top of its stroke. The space also may be hollowed out of the piston crown or cylinder head (see Figure 2-6). This is the simplest kind of combustion chamber and has a number of advantages.

Its surface area, relative to the volume of the combustion chamber, is less than in any other type of chamber. This means that less heat is lost to engine surfaces. As a result, the thermal efficiency is higher. It also makes starting easier, because less of the heat of compression dissipates to the cold engine. This feature allows lower compression ratios than those required with other types of chambers (often 16:1 as opposed to 20:1 or higher), leading to less stress on the engine and longer life.

In other types of combustion chambers, a portion of the air charge is forced in and out through small openings. This is often referred to as *work being done on the air*. The process generates a good deal of fric-

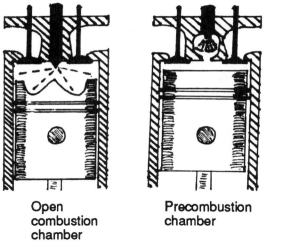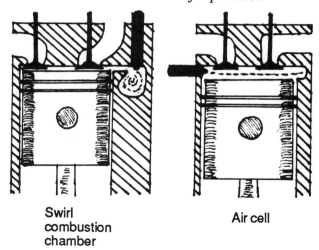

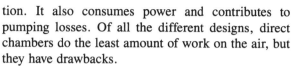

| Open combustion chamber | Precombustion chamber | Swirl combustion chamber | Air cell |

Figure 2-6. *Types of combustion chamber.*

tion. It also consumes power and contributes to pumping losses. Of all the different designs, direct chambers do the least amount of work on the air, but they have drawbacks.

A direct combustion chamber creates less turbulence than any other type; therefore, it uses less of the oxygen in the cylinder. For a given cylinder size, direct chambers generate less power than the others. In an attempt to offset this, inlet valves and seats are shaped and positioned in ways that impart a swirling

Figure 2-7. *Direct injection with a toroidal crown piston. (Courtesy Volvo Penta)*

motion to the air charge as it enters the cylinder. Further, the piston crowns on these engines are frequently given a convoluted shape (known as a *toroidal crown*, see Figure 2-7), which has the same effect. Currently there is a great deal of research being done on improving the efficiency of direct combustion chambers, with many manufacturers making a move back to this type of chamber.

Engines with direct combustion chambers almost always use hole-type injectors, which break fuel into smaller particles than pintle nozzles, thus aiding combustion.

Precombustion Chambers. Manufacturers often cast separate precombustion chambers into a cylinder head. These chambers range in size from 25% to 40% of the total compression volume of the cylinder. When fuel is injected into a precombustion chamber, it starts to burn, causing the temperature and pressure to rise above those in the main combustion chamber. This forces the unburned balance of the fuel and air mixture to rush through the precombustion chamber's relatively small opening into the main chamber, resulting in a high degree of turbulence and a thorough mixing of the fuel and air.

This type of engine generally uses pintle injectors, because their conical pattern distributes the diesel throughout the precombustion chamber. The extreme turbulence set up in the main combustion chamber offsets the pintle injector's reduced degree of atomiza-

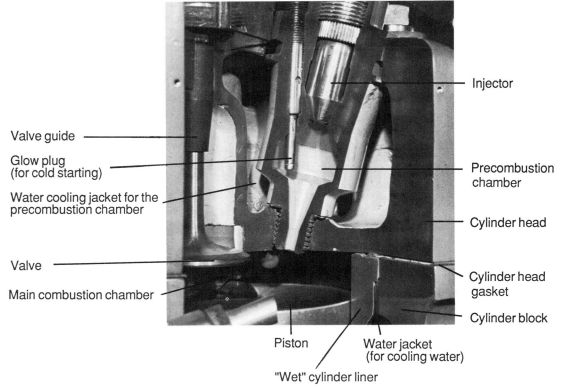

Valve guide

Glow plug
(for cold starting)

Water cooling jacket for the
precombustion chamber

Valve

Main combustion chamber

Injector

Precombustion
chamber

Cylinder head

Cylinder head
gasket

Cylinder block

Piston

Water jacket
(for cooling water)

"Wet" cylinder liner

Figure 2-8. *Cutaway view clearly shows a precombustion chamber. (Courtesy Caterpillar Tractor Co.)*

tion, compared with a hole-type nozzle.

Precombustion chambers make better use of the oxygen than direct chambers, resulting in more power from a given cylinder size. More work is done on the air, however, and the increased surface area of the two combustion chambers reduces thermal efficiency. Engine starting is harder due to the greater heat losses. For this reason, compression ratios are generally higher (from 20:1 to 23:1), and *glow plugs* (see Chapter 4) are invariably installed in the precombustion chambers to assist in cold starting (see Figure 2-8).

Swirl Chambers. Swirl chambers are similar to precombustion chambers, but their volume is almost equal to that of the main chamber. A very high degree of turbulence is imparted to the air charge as it enters the swirl chamber. Pintle nozzles inject the fuel into this swirling mass. Air use is high, but so too is the

amount of work done on the air. Thermal efficiency suffers, so compression ratios must be high. Glow plugs are needed for cold starting.

Other Variations. These three chamber types account for most injector nozzle/combustion chamber combinations found in marine diesels, but there are other variations. For example, *air cells* are sometimes used. In this arrangement an open chamber is set opposite the injector, and the fuel is sprayed across the piston top into the air cell, and combustion takes place throughout.

Despite the variety of combustion chambers and injectors employed, fuel and air never mix or burn 100%. For this reason, diesel engines are always designed to draw in more air than is strictly required to burn the fuel charge so that full combustion of the diesel is assured. The more complete the combustion, the more fuel-efficient and the less polluting the engine. Clearly, though, mixing fuel and air as thoroughly as possible reduces the amount of excess air needed for a clean burn, resulting in greater power from any cylinder of a given size.

At this point you may ask, "Why do I need to know these things?" The next time you see a glossy engine brochure advertising the merits of a "high-swirl combustion engine" you will know just what is being described. More important, the basis for effective troubleshooting is an understanding of what is going on inside your engine. For example, your engine has been increasingly difficult to start of late, has been emitting black exhaust smoke, and is now beginning to overheat and seize up while in use. You've determined that the cooling system is working fine, that the crankcase has plenty of oil, and that the oil pressure is normal (or perhaps marginally on the low side due to the overheating). You need to ask yourself a couple of questions: Does this engine have direct-chambers, precombustion-chambers, or swirl chambers? Does it have hole or pintle injectors?

If the engine has direct-combustion chambers with hole-type injectors, one or more injectors may be malfunctioning, leading to a poor atomization of the fuel and possibly to injector *dribble*. This could be the cause of the difficult starting and the black smoke from unburned fuel. The liquid fuel in the cylinder is now washing away the film of lubricating oil on part of a cylinder wall, creating excessive friction with the piston. The piston and cylinder are heating up, and a partial seizure is under way. Engines with precombustion and swirl chambers are less likely to suffer these symptoms.

This is only one possibility, and this is not the place to get into troubleshooting (this follows later). But this example does illustrate that you can never have too much information about an engine when diagnosing problems, and even the facts that seem most obscure can come in handy.

SECTION THREE: FUEL INJECTION

The first two sections of this chapter have described what must happen inside a cylinder if the injected fuel is to burn effectively. This section takes a look at the injection system itself, which is surely one of the miracles of modern technology. Today's breed of small high-speed, lightweight, and powerful diesels are possible in large part because of dramatic improvements in fuel-injection technology.

Consider a 4-cylinder, 4-cycle engine running at 3,000 rpm, and burning 2 gallons of diesel per hour. On every compression stroke, the fuel system will inject 0.0000055 gallons of fuel (5.5 millionths of a gallon!). Depending on the type of injector, injection pressures vary from 1,500 psi to 5,000 psi, so the fuel must be raised to this pressure, as well.

Each stroke of the piston of an engine running at this speed takes only $1/100$ of a second. In less than an instant, the injection system must initiate injection, continue it at a steady rate, then cut it off cleanly. The rate of injection must be precisely controlled. If it is too fast, combustion accelerates, leading to high temperatures and pressures in the cylinder and to engine *knocks* (for more on this see page 75). If it is too slow, combustion retards, leading to a loss of power and a smoky exhaust. The fuel, as we have seen, must be properly atomized and there must be no dribble from the tip of the injector, either before or after the injection pulse.

The actual beginning point of injection must be timed to an accuracy of better than 0.00006 seconds! Finally, every cylinder must receive exactly the same amount of fuel. It also must be constant from revolution to revolution to avoid vibration and uneven cylinder loading, which could lead to localized overheating and piston seizure.

These facts and figures serve only to illustrate that *a diesel engine fuel injection system is an incredibly precise piece of engineering that needs to be treated with a great deal of respect.*

The overwhelming majority of diesel engines use one of the following approaches to fuel injection: *jerk* pumps, a *distributor* pump, or a *modified common rail* system.

Jerk Pumps

A schematic of this type of fuel injection system is shown in Figure 2-9. A lift, or feed, pump draws the fuel from the tank, through a primary fuel filter. It then pushes fuel at low pressure through a secondary

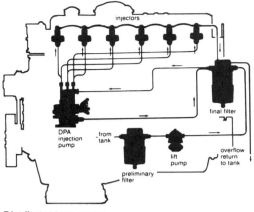

Distributor-type pump

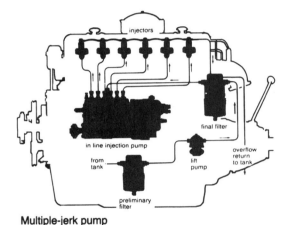

Multiple-jerk pump

Figure 2-9. *Schematics of jerk-pump (bottom) and distributor-pump (top) fuel injection systems. (Courtesy Lucas CAV Ltd.)*

filter to the jerk-type injection pump.

Jerk pumps consist of a plunger driven up and down a barrel by a camshaft. At the bottom of the plunger stroke, fuel enters the barrel. As the plunger moves upward, it forces this fuel out of the barrel through a check, or *delivery*, valve and to the injector (see Figure 2-10). The pressure generated by the pump forces open another valve in the injector, allowing injection to take place.

The speed of a diesel engine is regulated by controlling the amount of fuel injected into the cylinders. To make this possible, the jerk pump's plunger has a curved *spill* groove machined down its edge. A hole drilled from the top of the plunger connects with this groove so that the fuel in the barrel can flow to it (see

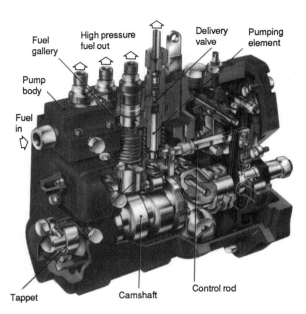

Figure 2-10. *Arrangement of plungers and barrels of an in-line jerk pump. (Courtesy Lucas CAV Ltd.)*

Figure 2-11). Another hole, a *bleed-off port*, is also drilled into the pump barrel.

Any time the spill groove lines up with the bleed-off port, the fuel in the barrel runs out, pressure falls, and injection ceases. By rotating either the barrel or the plunger, the groove and port can line up at a variety of points on the plunger's stroke, thereby varying the amount of fuel injected before the moment of bleed-off.

A gear fitted to the plunger or the barrel is turned via a geared rod, which is known as the *fuel rack*. The throttle is connected to this rack. Varying the position of the throttle regulates the flow of diesel to the injectors, which controls engine output.

Each cylinder requires its own jerk pump, but all of them are normally housed in a common block and driven by a common camshaft (with a separate cam for each pump). These are known as *in-line* pumps (see Figure 2-12).

Smooth engine operation from an in-line system demands that each pump put out exactly the same quantity of fuel to within *millionths* of a gallon. A jerk pump plunger has no piston rings to seal it in its barrel—it relies solely on the accuracy of the fit of the plunger and barrel. Nowadays, the two are machined to within 0.0004 ″ of each other, and plungers and bar-

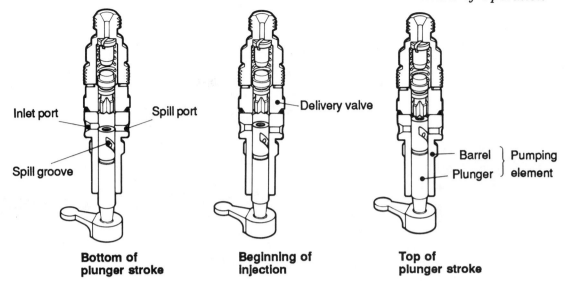

Figure 2-11. *The pumping element of a jerk pump. (Courtesy Lucas CAV Ltd.)*

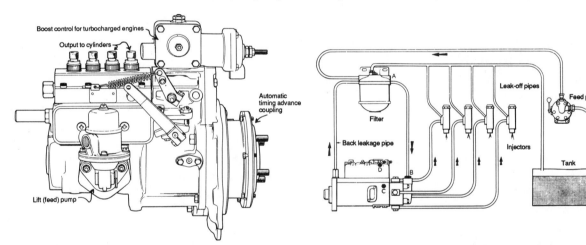

Figure 2-12. *A typical in-line jerk pump for a 4-cylinder engine. (Courtesy Lucas CAV Ltd.)*

Figure 2-13. *A typical fuel system with distributor pump. (Courtesy Lucas CAV Ltd.)*

rels must be as smooth as glass or they will seize up. Because of this degree of precision, amateurs should never tamper with fuel injection pumps. The only possible result is expensive damage!

Distributor Pumps

Figures 2-13 and 2-14 show this type of fuel-injection system. Note that it is broadly similar to the jerk

pump system, except for the addition of a *leak-off* pipe from the fuel injection pump back to the fuel tank. The two systems operate in the same way, use the same injectors, etc. The only difference between the two lies in the injection pumps themselves.

Whereas a jerk pump has a separate pump for each cylinder, a distributor pump uses one central pumping element and a rotating head that sends fuel to each cylinder in turn. This is done in much the same way a gasoline engine's distributor sends a spark to each spark plug in turn, hence the name *distributor*

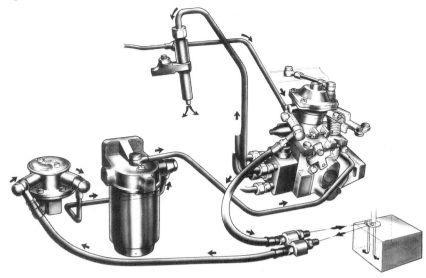

Figure 2-14. *Distributor-type fuel injection. (Courtesy Volvo Penta)*

pump. Because the same pump feeds all the cylinders, every injector gets an equal amount of fuel, ensuring even engine loading and smooth running at idle speeds. A metering valve on the inlet to the pump and connected to the throttle, regulates the output of the pump, therefore the engine (see Figure 2-15).

Common Rail Systems

Modified common rail units have a pump that draws fuel from a tank via a primary filter, passes the fuel through a secondary filter, and then discharges it continuously into a passage, or *gallery*, in the cylinder head. This gallery supplies the injectors (see Figure 2-16). A *pressure-regulating* valve at the end of the fuel gallery holds the pressure in the system at a set point and allows surplus fuel to return to the fuel tank. Fuel flows continuously through the whole system, including the injectors, keeping it lubricated and cool.

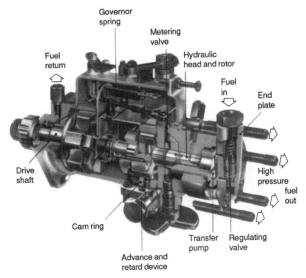

Figure 2-15. *Distributor-type fuel injection pump. (Courtesy Lucas CAV Ltd.)*

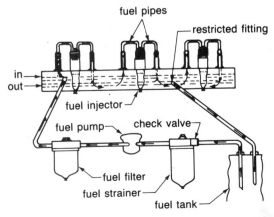

Figure 2-16. *Modified common-rail fuel injection system. (Courtesy Detroit Diesel Corp.)*

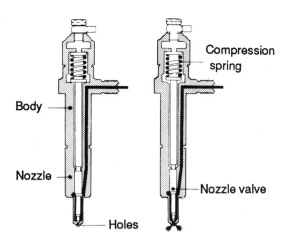

Body

Nozzle

Holes

Compression spring

Nozzle valve

Figure 2-17. Fuel injectors. Fuel entering the injector passes through galleries in the body and nozzle to a chamber surrounding the nozzle valve. The valve spring tightly holds the valve closed until fuel pressure from the injection stroke overcomes the spring's tension and lifts the valve, letting fuel under high pressure pass through the holes and tip of the spray nozzle. This happens instantaneously. At the end of the injection, fuel pressure rapidly falls and the spring returns the valve to its seat, ending the flow of fuel into the combustion chamber. (Courtesy Lucas CAV Ltd.)

Each injector contains its own fuel pump, similar in many ways to a jerk pump. An engine-driven camshaft operates each pump, and at the appropriate moment, the pump strokes and injects fuel directly into the engine.

Injectors

Every time an injection pump (of whatever type) strokes, it drives fuel at high pressure to an injector. Jerk pumps and distributor pumps pass fuel through a *delivery pipe* from the injection pump to the injector. A common rail system passes fuel to only the injection section of the combined pump/injector unit (known as a *unit injector*).

Inside the injector, a powerful spring holds a *needle valve* against a seat in the injector nozzle. The needle valve is designed to lift off its seat, allowing fuel to enter the engine, when the pressure in the

injector reaches a certain level (see Figure 2-17). On jerk pumps and distributor pumps, a small amount of diesel is allowed to find its way past the stem of the needle valve in order to lubricate the injector. The extra fuel then returns to the fuel tank via a leak-off pipe. Fuel continuously circulates through the body of the common rail system's injector.

Lift Pumps

All fuel systems use some kind of a *lift pump* to move diesel from the fuel tank to the injection pump. The modified common rail system requires a custom-made gear pump very specifically designed for this system. Other units need only a low-pressure pump suitable for use with diesel fuel. Some engines use electric diaphragm pumps similar to the units found on many automobiles, but the majority of jerk and distributor injection systems use a mechanically driven *diaphragm pump* fitted to the engine block or the side of the injection pump (for more on lift pumps, see page 65).

SECTION FOUR: GOVERNORS

The output of a diesel engine is controlled by regulating the amount of fuel injected into the cylinders. In marine use, you normally want the engine to run at a specific speed, regardless of the load placed on it. This cannot be done by simply pegging the throttle at a certain point because every time the load increases or decreases, the engine slows down or speeds up. Constant-running is achieved by connecting the fuel-control lever on the injection pump to a *governor*.

Simple Governors

The most basic governor consists of two steel weights, known as *flyweights*, attached to the ends of two hinged, spring-loaded arms (see Figure 2-18). The governor's drive shaft is mechanically driven by the engine, and as it spins, the flyweights spin with it, pushed outward by centrifugal force. A *speeder spring* counterbalances this centrifugal force.

Let us assume that the engine is running, the gov-

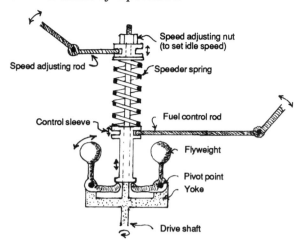

Figure 2-18. *A basic governor.*

ernor is spinning, and the flyweights are in equilibrium with the speeder spring at a certain position. If the load decreases and the engine speeds up, the governor will spin faster and the flyweights will move out under the increase in centrifugal force. As the flyweights move out, their arms push up against the control sleeve, which compresses the speeder spring until sufficient counterbalancing pressure restores equilibrium.

The control sleeve is connected by a series of rods to the injection pump's fuel-control lever, and when the sleeve moves up the governor's drive shaft, it cuts down the injection pump's rate of delivery. The reduction in fuel slows the engine to the speed at which it was originally set.

If the load increases and the engine slows down, the centrifugal force on the flyweights decreases, and they move inward under the pressure of the speeder spring. In moving inward, the flyweight arms allow the speeder spring to push the control sleeve down the drive shaft, operating the injection pump's fuel-control lever. This causes more fuel to be injected, which returns the engine to its preset speed.

The engine can be run at any set speed by adjusting the tension on the speeder spring via the speed-adjusting rod. The greater the pressure on the spring, the more the flyweights are held in, the more fuel is injected, and the faster the engine runs. Reducing the pressure on the speeder spring causes the flyweights to move out more easily, therefore sooner. When the

fuel injection rate is reduced, the engine runs at slower speeds.

On larger engines the simple mechanical governor just described is likely to be replaced by a complex hydraulic one, though the principles are the same. Some small marine diesels have a governor installed in the engine block, but building it into the back of the fuel-injection pump is becoming more common. It rarely malfunctions (troubleshooting is covered on page 104). Beyond the occasional need to adjust the tension of the speeder spring in order to set up the engine's idle speed, you should not need to know any more about your governor than the information given here.

Vacuum Governors

You may occasionally run across a vacuum-type governor. This operates as follows: *a butterfly valve* (a pivoting metal flap) is installed in the entrance to the air-inlet manifold. A vacuum line is connected between the air-inlet manifold and a housing on the back of the injection pump. Inside this housing is a diaphragm that is connected to the injection pump's fuel-control lever, or *rack*.

The throttle operates the butterfly valve, and when the throttle is shut down, the butterfly valve closes off the air inlet to the engine. The pumping effect of the pistons attempting to pull in air then creates a partial vacuum in the air-inlet manifold. This vacuum is transmitted to the housing on the fuel-injection pump via the vacuum line, sucking in the diaphragm against a spring. The diaphragm pulls the injection pump's control lever to the closed position.

If the load increases, the engine slows, the vacuum declines, and the spring in the fuel-injection pump pushes the diaphragm, and with it the fuel-control lever, to a higher setting. The engine speeds up. If the load decreases, the engine speeds up and the vacuum increases, sucking the diaphragm to a lower fuel setting and slowing the engine.

When the throttle is opened, the butterfly valve opens, the manifold vacuum declines, the diaphragm moves back under pressure from the spring, the fuel-control rack increases the fuel supply, and the engine speeds up to the new setting.

Aside from leaks in the vacuum line and around

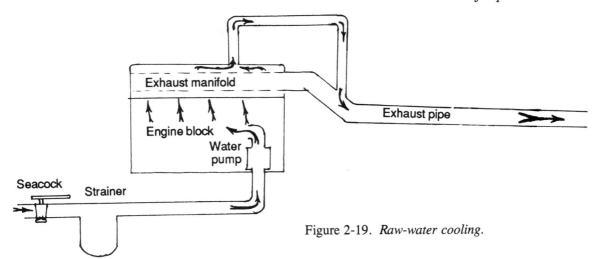

Figure 2-19. *Raw-water cooling.*

the diaphragm housing or a ruptured diaphragm (both are covered on page 105), little can go wrong with this system. The engine's idle speed is set with a screw that adjusts the minimum closed position of the butterfly valve.

SECTION FIVE: KEEPING THINGS COOL

Diesel engines generate a great deal of heat, only one-third of which is converted to useful work. The remaining two-thirds must somehow be released to the environment so that temperatures in the engine do not become dangerously high. Excessively high temperatures can break down lubricating oils, causing the engine to seize, or the cylinder head to crack. A little less than one-half of the wasted heat goes out with the exhaust gases; a similar amount is carried away by the cooling system. The balance radiates from the engine's hot surfaces.

Three principal types of cooling systems are used in boats: *raw water, heat exchangers*, and *keel coolers*.

Raw-water Cooling

Raw-water cooling systems draw directly from the body of water on which the boat floats. This water enters through a seacock, passes through a strainer, circulates through oil coolers, an aftercooler, and the

engine, and finally discharges overboard. On most 4-cycle diesels, after the water circulates through the engine, it enters the exhaust pipe and discharges with the exhaust gases. This is a *wet* exhaust (see Figure 2-19); a *dry* exhaust has no water injection.

A raw-water cooling system is simple and economical to install, but it has a number of drawbacks.

- Over time a certain amount of silt inevitably finds its way into the engine and coolers and begins to plug cooling passages.
- Regulating the engine's temperature is difficult. First, the temperature of the raw water may range from the freezing point in northern climates in the winter to 90°F (32°C) in tropical climates in the summer. Second, the system frequently has no thermostat because the inevitable bits of trash and silt can clog it and make it malfunction.
- In salt water, scale (or salts) crystallize out in the hottest parts of the cooling system, notably around cylinder walls and in cylinder heads. This leads to a reduction in cooling efficiency and localized hot spots (General Motors states that 1/16″ of scale on 1″ of cast iron has the insulating effect of 4 1/2″ of cast iron!). The rate of scale formation is related to the temperature of the water, and accelerates when coolant temperatures are above 160°F (71°C). As a result, raw-water-cooled engines generally run cooler than other engines (around 140°F to 160°F [60°C to 71°C], compared with 185°F

Side of block
split out

Figure 2-20. *Cracked block on a raw-water-cooled engine. This engine was mounted below the waterline and had no siphon breaks. When it froze, the block split. When it thawed, water flowed in through the cracked block and sank the boat.*

[85 °C] or higher). This lowers the overall thermal efficiency of the engine and can also cause water and harmful acids to condense out in the engine oil (see page 34).

- The combination of heat, salt water, and dissimilar metals is a potent one for galvanic corrosion. An engine using raw-water cooling should be made of galvanically compatible materials (e.g., a cast-iron block should have a cast-iron cylinder head, not an aluminum one). Sacrificial zinc anodes must be installed in the cooling circuit, and they have to be inspected and changed regularly (see page 92).

- The engine cannot be protected with antifreeze in cold weather (see Figure 2-20); therefore, it will have to be drained after every use. All piping runs and pumps need to be installed without low spots in order to facilitate draining.

Today's powerful, high-speed, lightweight diesels demand a greater degree of cooling efficiency than that provided by raw-water cooling. Almost all of these power plants use a heat exchanger or keel cooler.

Heat Exchanger Cooling

An engine with a heat exchanger has an enclosed cooling system. The engine cooling pump circulates the coolant from a header tank through oil coolers, an aftercooler, and the engine. The coolant then passes through a heat exchanger to lower its temperature and is circulated once again (see Figure 2-21).

A heat exchanger consists of a cylinder with a number of small cupronickel tubes running through it (see Figure 2-22). The hot engine coolant passes through the cylinder while cold raw water is pumped through the tubes. The raw water carries off the heat from the coolant then discharges overboard, directly or via a wet exhaust.

A heat exchanger is expensive, requires more piping than a raw-water system, and an extra pump on the exchanger's raw-water circuit. On the other hand, antifreeze and corrosion inhibitors can be added to the coolant to protect the engine against freezing and corrosion. No silt finds its way into the engine, and no salts crystallize out of the coolant around cylinder liners and in the cylinder head. The engine can be operated at higher temperatures, which is more thermally efficient. The expansion tank will have a pressurized cap, as on an automobile's radiator. When the pressure of the coolant is increased, so too is its boiling point. If the pressure is raised by 10 psi, its boiling point rises to approximately 240 °F (116 °C). Allowing the pressure to rise in a closed cooling system greatly reduces the risk of localized pockets of steam forming and causing damage at hot spots in the engine.

The raw-water side of a heat-exchanger circuit will still suffer from many of the problems associated with raw-water cooling, particularly silting up of the heat-exchanger tubes, and the potential for damage from corrosion and freezing. Sacrificial zincs are

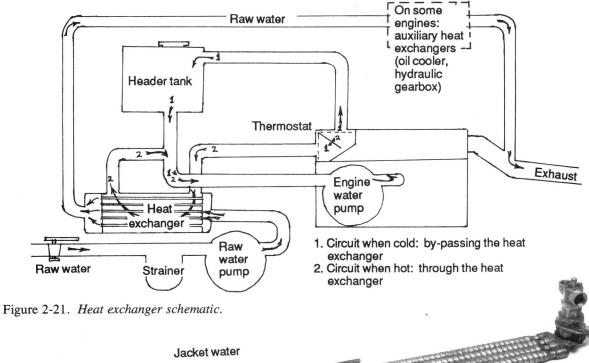

Figure 2-21. *Heat exchanger schematic.*

1. Circuit when cold: by-passing the heat exchanger
2. Circuit when hot: through the heat exchanger

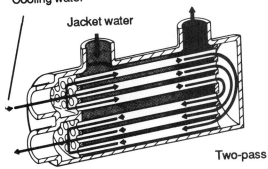

Figure 2-22. *Heat exchangers. (Courtesy Caterpillar Tractor Co.)*

Figure 2-23. *A keel cooler. (Courtesy The Walters Machine Co. Inc.)*

essential once again, and the piping must have no low spots so that it can be easily drained in cold weather.

Keel Cooling

Keel cooling does not require a raw-water circuit. Instead of placing a heat exchanger in the boat and bringing raw water to it, a heat exchanger placed outside the boat is immersed directly in the raw water (see Figure 2-23). This is normally done by running a pipe around the keel of the boat and circulating the engine coolant through it, or by installing an arrangement of cooling pipes on the outside of the hull. Steel boats sometimes have channel-iron passages welded to the outside, or a double skin with the engine coolant circulating between the two skins. Neither of these is recommended because each adds a good deal of weight, and repairs can be very expensive if the channel iron or either skin rusts or gets damaged.

A keel cooler has all the advantages of a heat exchanger plus a few more. It does not suffer any problems with silting, corrosion, or freezing on any part of the cooling circuit. If the engine has a dry exhaust, it does not need any kind of a raw-water pump or circuit. If the engine has a wet exhaust, a separate raw-water pump is necessary to supply the water injected into the exhaust. This pump often also circulates water through an aftercooler, and perhaps the oil cooler for the engine or transmission. The best keel coolers are made of bronze with cupronickel tubing.

Wet and Dry Exhausts

Exhaust gases exit cylinders at very high temperatures and in considerable volume. They must be removed from the boat as efficiently as possible. Any pressure

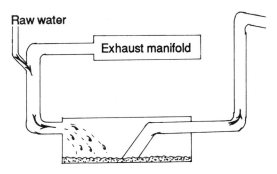

Figure 2-24. *This exhaust is cooled and silenced by water.*

build-up in the exhaust system (*back pressure*) interferes with the smooth flow of gases through the engine and reduces performance. This is another component of the pumping losses discussed earlier. Although a large-diameter straight exhaust pipe will remove the gases with as little interference as possible, the noise level is totally unacceptable.

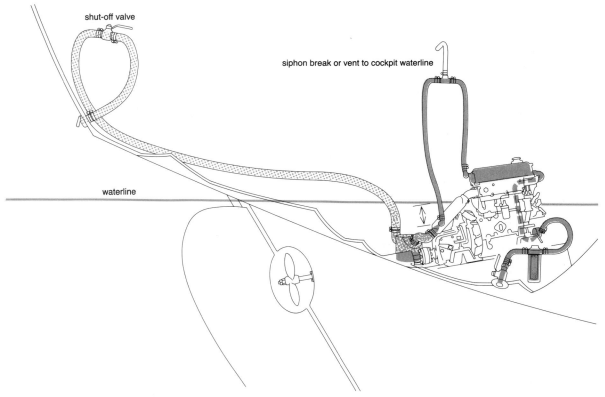

Figure 2-25. *Overview of the exhaust system.*

Noise is a rather complicated business, but one of its major causes is the velocity with which gases exit an engine. Another is the sudden pressure changes created as each cylinder discharges its exhaust gases. Decreasing the volume of the gases or expanding them into a larger area reduces velocity. A certain amount of back pressure in the exhaust system smooths out pressure changes.

This is where a wet exhaust comes into its own (see Figure 2-24). The high temperature of the exhaust gases causes some of the water injected into the system to vaporize (boil). In vaporizing, the water absorbs a good deal of heat from the exhaust (latent heat of vaporization), sharply lowering the temperature. The fall in temperature produces a corresponding decrease in exhaust gas volume, and this reduces the velocity. A wet exhaust cools the gases and partially silences the system with no increase in back pressure. What is more, beyond the water injection point a suitable rubber hose can be used for the exhaust pipe. This absorbs some of the noise that would radiate from the metal pipes necessary to withstand the heat of dry exhausts.

Silencing in a wet exhaust can be further improved with only a small increase in back pressure by using a water-lift-type silencer. Here's how it works. After the water is injected into the exhaust, the water and gases flow into an expansion chamber, which has an exit near its base. The unvaporized water builds up until it blocks this exit, at which point the exhaust gases blow it out of the exhaust pipe (see Figure 2-25). The expansion chamber, and small degree of back pressure, combine to even out the pressure changes in the exhaust system and so reduce noise.

Although wet exhausts are very effective and increasingly popular, several precautions regarding their installation must be observed or water may find its way into the engine. This has been the ruination of many an engine and has even caused the sinking of a few boats. These precautions are covered in Chapter 9.

Two-Cycle Engines

The exhaust gases of a 4-cycle engine are pushed out in part by the piston on its fourth stroke. High exhaust back pressure will not, therefore, prevent the removal of the gases, but it will cause the engine to work harder, run hotter, and lose power. The exhaust gases of a 2-cycle diesel, however, are cleaned out by the pressure of the scavenging air blown into the cylinder. A high enough exhaust back pressure will completely stall this flow, and the engine simply will not run.

For this reason, most 2-cycle diesels have fairly direct dry exhausts that are considerably noisier than their 4-cycle counterparts. This noise is exacerbated by the relatively short duration of the exhaust cycle, which requires the exhaust valves to open and close more rapidly, creating greater pressure changes in the exhaust gases. There is a limit to what can be done to muffle this racket without affecting the engine's performance. (The most effective silencing will probably be achieved with the "modified" wet silencer described in Chapter 9, but even so the engine will still be relatively noisy.)

Dry exhausts run much hotter than wet exhausts. They require thorough insulation, especially where they pass through bulkheads or where there is a danger of contact with the crew.

Chapter 3

Cleanliness Is Next to Godliness

Diesel engines are remarkably long-lived and reliable—these are two of their principal attractions. What is more, they require little routine maintenance, but *the little maintenance they do need is absolutely essential*. Carelessness and inattention to detail can lead to thousands of dollars of damage in as little as a few seconds.

A diesel engine must have clean air, clean fuel, clean oil, and be kept clean.

If this book does no more than provide an understanding of why cleanliness is so important, and instill in the reader a determination to change air, fuel, and oil filters at the specified maintenance intervals, it will have been a success.

Clean Air

As we have seen, about 1,500 cubic feet of air at 60 °F (15.6 °C) is required to completely burn one gallon of diesel. But even this figure considerably understates the actual amount af air used by a diesel engine. The pistons of a naturally aspirated engine pull in about the same amount of air at each inlet stroke, regardless of engine speed or load (on supercharged and turbocharged engines the air intake will vary with speed and load). At low speeds and loads, very little fuel is

injected, and the oxygen in the cylinder is only partly burned. As the load or speed or both increase, more fuel is injected, until at full load enough fuel is injected to burn all the oxygen (including the pressurized air from a supercharger or turbocharger). When the engine consumes all of the oxygen, it is at its maximum power output. (In practice, however, the maximum fuel injection is generally kept to a level at which only 70% to 80% of the oxygen is burned, in order to ensure complete combustion and keep down harmful exhaust emissions.)

Thus, at light loads, only a small proportion of the air drawn into an engine is burned; even at full loads there is a margin of unburned air to ensure complete combustion of the diesel and to keep exhaust pollutants to a minimum. Turbocharged engines use twice as much air as naturally aspirated models, and 2-cycles as much as four times more. This adds up to an awful lot of air. (Figure 3-1 gives some idea of the volume of air used by a naturally aspirated 4-cycle diesel.)

The efficient running of a diesel absolutely depends on its maintaining compression. Even small amounts of fine dust passing through a ruptured air filter or a leaking air-inlet manifold leads to rapid piston-ring wear and scoring of cylinders, which pave the way to expensive repairs. What is more, *once dirt*

28

	The volume of air required by a naturally aspirated 4-cycle engine running at 83% volumetric efficiency.						
CID*/liters	Engine speed (RPM)						
	500	1000	1500	2000	2500	3000	Cubic feet per minute
50/0.8	6	12	18	24	30	36	
75/1.25	9	18	27	36	45	54	
100/1.6	12	24	36	48	60	72	
125/2.0	15	30	45	60	75	90	
150/2.5	18	36	54	72	90	108	

*CID = Cubic Inches of Displacement

These figures are calculated with the following formula: $\dfrac{CID \times (^1/_2 \text{ engine speed}) \times 0.83}{12 \times 12 \times 12}$

Figure 3-1. *Air consumption table.*

gets into an engine, properly cleaning it out is impossible. Small particles become embedded in the relatively soft surfaces of pistons and bearings, and no amount of oil changing and flushing will break them loose. This dirt accelerates wear.

Every hour, day after day, sometimes year after year, a small diesel sucks in enough air to fill a large room, yet as little as two tablespoons of dust contained in that air can do enough damage to necessitate a major overhaul.

Even if a filter is not ruptured, as it filters the air it becomes plugged, progressively restricting airflow to the engine. This limits the amount of oxygen reaching the cylinders, and combustion, especially at higher loads, suffers. The engine begins to lose power, and the exhaust shows black smoke from improperly burned fuel. Valves, exhaust passages, and turbochargers carbon up (as will the engine itself), further reducing efficiency and leading to other problems (see below for the effects of carbon in the oil). The engine is likely to overheat, and in extreme cases seize.

Air filters must be kept clean! The interval for changing air filters, however, depends on the operating conditions. In general, the marine environment is relatively free of airborne pollutants, making air filter changes an infrequent occurrence. Since this can lead to complacency and a forgotten air filter, changing the filter at a set interval, even if it appears to be clean, is far better than forgetting it.

Figure 3-2. *Air filter with replaceable paper element. (Courtesy Caterpillar Tractor Co.)*

Air Filters. Air filters on most small diesels are of the replaceable paper-element type found in automobiles (see Figure 3-2). Less common is the oil-bath type (see Figure 3-3). The latter type forces the air to make a rapid change of direction over a reservoir of oil. Particles of dirt are thrown out by centrifugal force and trapped in the oil. The air then passes through a fine screen, which depends on an oil mist drawn up from the reservoir to keep it lubricated and effective.

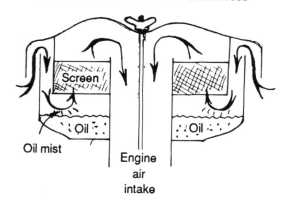

Figure 3-3. *An oil-bath air cleaner.*

In time, although the oil may still look clean, the reservoir fills with dirt, the oil becomes more viscous, less oil mist is drawn up, and the filter's efficiency slowly declines. The oil must periodically be emptied from the reservoir and the pan thoroughly cleaned with diesel or kerosene. At this time, the screen should also be flushed with diesel or kerosene and blown dry. When refilling the reservoir with oil, *be careful not to overfill it*—excess oil can be sucked into the engine causing damaging *detonation* and *run-away* (for more on these see page 79).

Clean Fuel

A fuel-injection pump is an incredibly precise piece of equipment, which can be disabled by even *micro-scopic* pieces of dirt or traces of water. It is also *the single most expensive component on an engine*, and about the only one that is strictly off-limits to the amateur mechanic. Attempts to solve problems invariably make matters worse. It is therefore of vital importance to be *absolutely fanatical about keeping the fuel clean*. Yet so many boatowners treat their fuel systems with indifference. According to CAV, one of the world's largest manufacturers of fuel injection equipment, the result is that 90% of diesel engine problems result from contaminated fuel.

Fuel is contaminated by dirt, water, and bacteria. Even minute particles of dirt can lead to the seizure of injection-pump plungers, or to scoring of the cylin-

ders and plungers. Scoring in jerk pumps allows fuel to leak by the plungers, resulting in uneven fuel distribution. Even variations of a few millionths of a gallon in the amount of fuel injected into each cylinder adversely affects performance, causing rough running, uneven loading from one cylinder of the engine to another, and a loss of performance. If the dirt finds its way to the injectors themselves it can cause a variety of equally damaging problems, such as plugged or worn injector nozzles.

If an engine runs unevenly, the cylinders that carry the extra load are likely to overheat, which may cause burned valves and pistons, a cracked cylinder head, or even a complete seizure. Worn or damaged injectors tend to dribble, resulting in a smoky (black) exhaust. In some instances, misdirected or improperly atomized streams of fuel will wash the film of lubricating oil off a section of a cylinder wall, resulting in the seizure of that piston.

Water in the fuel opens another can of worms. Aside from causing misfiring and generally lowering performance, water droplets in an injector can turn to steam in the high temperatures of a cylinder under compression. This happens with explosive force, which can blow the tip clean off an injector. Raw fuel is then dumped into the cylinder, washing out the film of lubricating oil while the injector tip rattles around, beating up the piston and valves. During extended periods of shut-down, which are quite common with most boat engines, water in the fuel system will also cause rust to form on many of the critical parts.

Bacteria grow in even apparently clean diesel, creating a slimy, smelly film that plugs filters, pumps, and injectors. The microbes live in the fuel/water interface, requiring both liquids to survive. They find excellent growth conditions in the dark, quiet, non-turbulent environment found in most fuel tanks. Two types of biocide are available to kill these bacteria. The first is water-soluble, the second diesel-soluble; the latter is preferred. Follow the instructions on the can when you add these chemicals to a tank.

Various other diesel-fuel treatments on the market are not generally recommended by fuel injection specialists. Some, for example, contain alcohol (to absorb water), but this attacks O-rings and other non-metallic parts in some fuel system equipment. Rather than treat your fuel after you have a problem, you should try to forestall any problems at source by doing the following:

- Ensure that all cans used for carrying fuel are spotlessly clean.
- When taking on fuel from a barrel, first insert a length of clear plastic tubing to the bottom of the barrel, plug off the outer end with your finger, and then withdraw the tube. It will bring up a sample of fuel from all levels of the barrel, indicating whether it is seriously contaminated.
- Filter all fuel, using a funnel with a fine mesh or preferably one of the multi-stage filter funnels now available through various marine catalogs and at some marine chandleries. If you detect any signs of contamination, stop refueling at once.
- Take regular samples from the bottom of your fuel tank to check for contamination. If you can't get at a drain valve, find some means of pumping out a sample of fuel. At the first sign of contamination, drain the tank or pump out the fuel until *no trace* of contamination remains. Any especially dirty batch of fuel should be *completely discarded*—it's not worth risking the engine for the sake of a tankful of fuel.
- When leaving the boat unused for long periods (e.g., when laid up over the winter), fill the fuel tank to the top. This eliminates the air space and cuts down on condensation of moisture in the tank. Add a biocide.

Fuel Filters. For all its precision, diesel fuel—injection equipment is heavily constructed. Diesel fuel is a lubricant, and since all working parts of the system are permanently immersed in it, friction and wear are virtually absent. If a fuel injection system has a constant supply of clean fuel it should give thousands of hours of trouble-free use. (Injector nozzles are the only exception, because they are subjected to such intense operating conditions in the combustion chambers. Injectors need to be pulled and cleaned at more frequent intervals—say every 900 hours.) Apart from ensuring that you take on clean fuel, all the fuel system really needs is a routine attention to the filters.

Without exception, every marine diesel engine should have both a primary and a secondary fuel filter. All engines come from the manufacturer with an engine-mounted secondary filter located some-

CAV watertrap

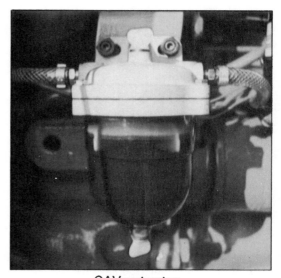

CAV waterstop

Figure 3-4. *Primary fuel filters. Fuel enters the CAV Waterscan (see next page), passes over and around the sedimenter cone, through the narrow gap between the cone and the body, then to the center of the unit and out through the head and outlet connections. This radial flow allows gravity to separate from the fuel water and heavy abrasive particles, which settle at the bottom of the bowl. This filter doesn't have any moving parts. The electronic probe in the base of the filter* (continued on page 32) *contains two electrodes; the filter itself is the third electrode. As the level of water increases, it disturbs the balance in the system and triggers a warning that the time has come to drain the bowl. The warning can be a light, buzzer, or other device. Removing a thumbscrew opens the drain hole. The unit has an automatic circuit that triggers the warning device for a period of 2 to 4 seconds when the system is first energized. (Courtesy Lucas CAV Ltd.)*

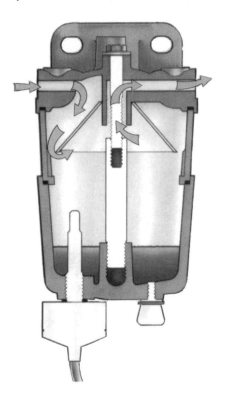

where just before the fuel injection pump. If this is the only filter, a primary filter MUST be installed. This needs to be mounted between the fuel tank and the lift pump, not *after* the lift pump, because any water in the fuel supply that passes through a lift pump gets broken up into small droplets that are hard to filter out.

Primary and secondary filters do not have the same function. A primary filter is the engine's main line of defense against water and serious contamination of the fuel supply, but it does not guard against microscopic particles of dirt and water. These are filtered out by the secondary filter.

A primary filter needs to be of the *sedimenter* type specifically designed to separate water from fuel. Sedimenters are extremely simple, generally consisting of little more than a bowl and deflector plate. The incoming fuel hits the deflector plate then flows around and under it to the filter outlet. Water droplets and large particles of dirt *settle out* and are *jetted out* by centrifugal force (see Figure 3-4). The better quality filters then pass the fuel through a relatively coarse filter element (10 to 30 microns—a micron is one millionth of a meter, or approximately 0.00004″—see Figure 3-5).

A primary filter should have either a see-through bowl with a drain so that water can be rapidly detected and removed, or a drain petcock so that a sample can be taken from the filter at regular intervals. Beyond this, the filter may have an electronic sensing device that sounds an alarm if water reaches a certain level, a float device that shuts off the flow of fuel to the engine if the water once again reaches a certain level, or both.

Powerboats should have two or more primary filters mounted on a valved manifold that allows either filter to be closed off and changed without shutting down the engine (see Figure 3-6). This way, if you have a problem with dirty fuel but are in a situation that makes shutting down the engine dangerous, you can change the filters while the engine runs. Such an arrangement would also be a good idea on many motorsailers. A vacuum gauge mounted between the primary filters and the lift pump is also an excellent troubleshooting investment. Any time the filters start to plug up, the rising vacuum will alert you.

Secondary filters are designed to remove very small particles of dirt and water droplets. They cannot handle major contamination because their fine mesh will soon plug up. Secondary filters are normally of the spin-on type and contain a very fine specially impregnated paper element that catches dirt. Water droplets are also too large to pass through this mesh and therefore adhere to it. As more water is caught, the droplets increase in size (*coalesce or agglomerate*) until they are large enough to settle to the bottom of the filter, from where they can be periodically drained. The filter mesh should be in the range of 7 to 10 microns, but certainly no larger (see Figure 3-7).

Lift pumps also normally have a fine screen on the inlet side to filter out large particles of dirt. If you find

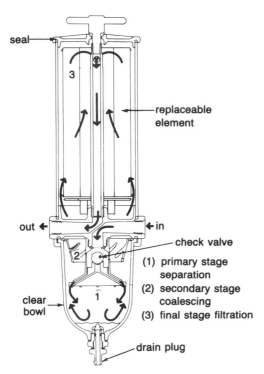

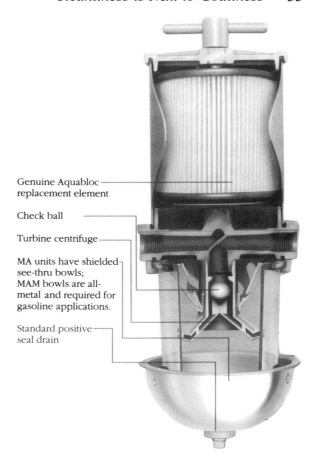

Figure 3-5. *Combined primary and secondary filter. (Courtesy RACOR)*

Figure 3-6. *These primary filters have valves that allow any one of them to be changed while the engine continues to run. (Courtesy RACOR)*

evidence of serious contamination in a primary filter, this screen should be checked. On engine-driven diaphragm-type pumps (the majority), the screen is accessible by removing a screw in the center of the pump cover (see Figure 3-8).

Regular fuel filter changes must be at the top of any maintenance schedule (for filter change procedures, see below). Just to drive home the point one more time, here's the true-life story of a friend of ours. He took on a batch of dirty fuel but failed to notice it. The contamination overwhelmed his primary filter and plugged the secondary filter. The secondary filter collapsed under the suction pressure from the injection pump and flooded the injection pump and injectors with a mass of dirty particles. The engine continued to run long enough to destroy the injection pump. Meanwhile, the messed up injection patterns caused by the dirt resulted in the overheating

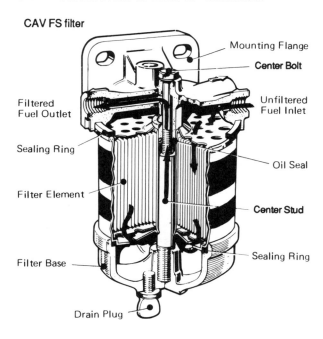

CAV FS filter

Mounting Flange

Center Bolt

Filtered Fuel Outlet

Unfiltered Fuel Inlet

Sealing Ring

Oil Seal

Filter Element

Center Stud

Filter Base

Sealing Ring

Drain Plug

Figure 3-7. Secondary filter with "agglomeration" capability. (Courtesy Lucas CAV Ltd.)

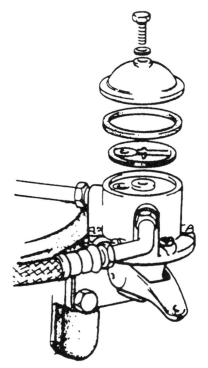

Figure 3-8. Cleaning the lift pump. (Courtesy Perkins Engines Ltd.)

and finally the seizure of two cylinders. Repairing the engine would have been more expensive than the cost of a new engine, so he bought a new engine.

Clean Oil

Lubricating oil in a diesel engine works much harder than that in a gasoline engine, owing to the higher temperatures and greater loads encountered. This is especially the case with today's lightweight, high-speed, turbocharged diesels, which contain a considerably smaller volume of oil than is found in traditional engines.

What's more, diesel fuels contain traces of sulphur, which form sulfuric acid when they mix with the water that is a normal by-product of the combustion process. Carbon (soot) is another by-product of combustion. The presence of carbon is the reason for the black color of diesel engine oil after just a few hours of engine running.

Boats traveling overseas often take on inferior fuel, the most damaging component of which is higher sulfur levels. Many cruising boats, particularly auxiliary sailboats, compound problems with inferior fuels by repeatedly running their engines at light loads to charge the batteries and run the refrigeration at anchor. The engines spend much of their lives running cool, which increases the amount of moisture condensing out in the engine. These condensates combine with the higher sulfur traces to make acid, which attacks sensitive engine surfaces. Low-load and cool running also generate far more carbon than normal. This gums up piston rings and coats valves and valve stems, leading to a loss of compression and numerous other problems. *Low-load, and low-temperature running is a thoroughly destructive practice*, about which I shall have more to say later.

Diesel engine oils are specially formulated to hold soot in suspension and deal with acids and other harmful by-products of the combustion process. Using the correct oil in a diesel engine is vitally important. Many perfectly good oils designed for gasoline engines are not suitable for use in a diesel engine. The American Petroleum Institute (API) uses the letter "C" (for compression ignition) to designate oils rated for use in diesel engines, and the letter "S" (for spark ignition) to designate oils rated for use in gasoline engines. The C or S are then followed by another letter to indicate the complexity of the additive package in the oil, with the better packages being given a letter further into the alphabet. Thus any oil rated CC, CD, or CE is suitable for use in diesel engines, with the CE oil being the best. In 1991 a new classification—CF-4—was added, with a further addition (tentatively labelled "CX") planned for 1994. Cruisers going to Third World countries should carry a good stock of the best grade of oil that money can buy.

As the oil does its work, the additives and detergents are steadily used up. The oil wears out. It must be replaced at frequent intervals—far more frequently than in gasoline engines. In particular, if high-sulfur content fuels are taken on, such as are likely to be found in many Third World countries and much of the Caribbean, oil change intervals should be shortened. Every time you change the oil, you must install a new filter to rid the engine of all its contaminants.

If regular oil changes are not carried out, sooner or later the acids formed will start to attack sensitive engine surfaces. The carbon will overwhelm the detergents in the oil, forming a thick black sludge in the crankcase and in the oil cooler (if fitted). The sludge will begin to plug narrow oil passages and areas through which the oil moves slowly, eventually causing a loss of supply to some part of the engine. A

Dirt	43%
Lack of oil	15%
Misassembly	13%
Misalignment	10%
Overloading	9%
Corrosion	5%
Other	5%
	100%

Figure 3-9. *Major causes of bearing failure.*

major mechanical breakdown is under way—and all for the sake of a gallon or so of oil, a filter, and less than an hour's work. One major bearing manufacturer estimates that 58% of all bearing failures are the result of dirty oil, or a lack of oil (see Figure 3-9). Oil sludge in oil coolers is almost impossible to remove and generally necessitates a new cooler.

Changing engine oil is sometimes complicated by the location of the drain plug in a place that you can't reach. Even if you can reach the plug, there isn't room enough to slide a container under the engine to catch the old oil. It may be possible to slip a piece of tubing into the dipstick hole, attach a pump to it, and remove the oil this way, but a better arrangement is to fit the engine with a small sump pump. This may be electrically driven, but generally a hand pump is more than adequate.

A copper pipe or hose fitting normally is screwed into the drain in the oil sump, and the pump plumbed into this pipe or fitting. Remember that if the pipe, hose, or connections fail, *you will suffer a catastrophic loss of engine oil*, so make sure the installation is to the highest standards. If you have an electric pump, there must be absolutely no way it can be turned on accidentally! Any pump will have to be compatible with engine oil. Change only hot oil; hot oil has lower viscosity, making it easier to pump and ensuring that all of it drains down out of the engine.

Oil analysis can reveal all kinds of trouble in the making and suggest ways to stave off problems. Once a year, save a sample of oil when you're changing it and send this to a laboratory for a spectroscopic analysis. You should be able to find the address of a local laboratory in the yellow pages; look under "laboratories - spectrographic," or check with any large machine shop or engine overhaul facility.

Figure 3-10. *Cleaning a watertrap filter. Before you begin, clean off all external dirt. If the sedimenter uses a gravity feed supply, turn off the fuel before you dismantle the unit. Slacken off the thumbscrew in the base and drain the water and sludge. (1) Unscrew the center bolt while you hold the base to prevent it from rotating. (2) Detach the base and separate it from the sedimenter element. Inspect the center sealing ring for damage and replace it if it's not perfect. (3) Clean the base and rinse it with clean diesel fuel. Do the same to the sedimenter element. (4) Clean out the sedimenter head and inspect the upper sealing ring for damage. Replace it with a new one it it's not per-fect. You can buy the sealing ring from the supplier who sold the filter element to you. (5) Make sure that the center sealing ring is correctly positioned and place the sedimenter element-cone pointing upward-onto the base. (6) Be sure that the upper sealing ring is correctly placed in the head, then install the assembled element and base. (7) Engage the center bolt with the central tube and make sure the top rim of the sedimenter element seats correctly before you tighten the bolt to a torque of 6 to 8 lbs. ft. (0.830 to 1.106 kg. m.). Do not overtighten the center bolt in an attempt to stop leaks. Tighten the drain thumbscrew only hand tight.*

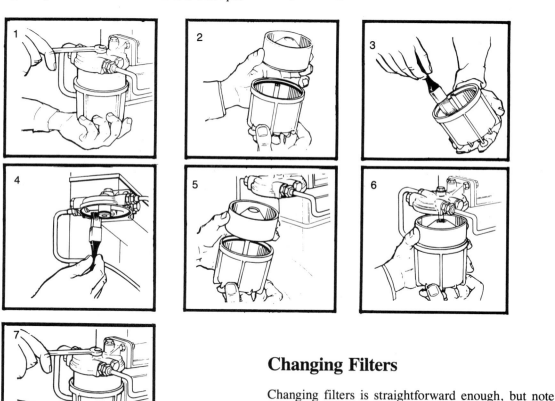

Changing Filters

Changing filters is straightforward enough, but note that most diesel fuel systems need *bleeding* after a filter change (see page 58). First, scrupulously clean off any dirt from around the old filter or filter housing (see Figure 3-10). Next, provide some means to catch any fuel or oil spilled—I find disposable diapers (especially the ones with elastic sides since they can be formed into a bowl shape) to be ideal. Most primary fuel filters have a central bolt or wing nut that you loosen to drop the filter bowl. If the filter has a replaceable element, take a close look at the old one.

Finally, please don't dump your old oil overboard or down the nearest drain. Save it in some old oil or milk jugs and take it to a proper disposal facility. If you don't know of one, ask at a local garage or call your state department of environmental protection.

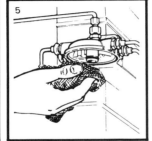
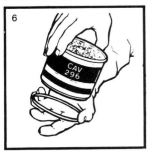

Figure 3-11. *Changing a filter element with a replaceable element. (1) Clean off all external dirt before you service the filter. Unscrew the thumbscrew in the base and drain water and sludge. (2) Unscrew the center bolt while you hold the base to prevent it from turning. (3) Release the filter element complete with the base by pulling the element downward while you turn it slightly so that it comes free of the O-ring. (4) Detach and discard the element. Detach and inspect the lower sealing ring for damage. Replace a defective O-ring. (5) Clean out the sedimenter base. Complete the cleaning by rinsing with clean diesel fuel. Clean the head and inspect the upper sealing ring and the O-ring for damage. Replace defective sealing rings. (6) New sealing rings are available from filter suppliers. (7) Make sure that the upper sealing ring and O-ring are positioned correctly in the head and fit a new filter element to the head. Rotate the element slightly when you fit it so that it slides easily over the O-ring. Ensure that the lower sealing ring is correctly placed in the base then install the base onto the assembled head and element. Make sure that the rims of the element and base seat correctly before you tighten the center bolt. Do not overtighten. (Courtesy Lucas CAV Ltd.)*

If it isn't spotless—as it should be in a well-maintained fuel system—find out where the contamination is coming from and stop it before it stops you.

Screw-on filters (both fuel and oil—see Figure 3-11) are undone with an appropriate filter wrench. This should be a part of the boat's tool kit; note that more than one size of filter wrench may be needed for fuel and oil filters. In the absence of a filter wrench, wrap a V-belt around the filter, grip it tightly, and unscrew it. This generally will provide enough leverage to loosen it. Failing this, you can always hammer a large screwdriver through the filter—it may be messy, but at least it will enable you to get the filter off.

Fuel filters are often filled with clean diesel before installation. This reduces the amount of priming and bleeding that has to be done, but this practice carries with it the possibility of contaminants being introduced directly into the injection system. For this reason, *never fill the secondary filter before installing it*. Normally the priming can be done by operating the lift pump manually (see page 59) but sometimes on larger installations it pays to install an additional electrically-operated lift pump on a separate by-pass manifold. This pump is placed before any filters and is used to push fuel through the filters, priming the system.

If the new filter has its own sealing ring, ensure that the old one doesn't remain stuck to the filter housing. If the new filter has no sealing ring, you will have to use the old one. To prevent this, buy a stock of rings and fit one at each filter change. Note that some sealing rings have a square cross section—they must go in without twisting. The sealing rings of screw-on filters should be lightly lubricated before installation. These filters are done up hand tight, and then given an additional three-quarters of a turn with the filter wrench.

If the filter is done up with a wing nut, check closely for leaks around the nut when you are finished. This is a likely source of air in a fuel system and one of the first places to look if a running problem arises immediately after a filter change.

Note that some turbochargers have their own oil filter. This must be replaced whenever the engine oil filter is changed.

Routine oil and filter changes take little time and are relatively inexpensive but are too often neglected. Remember, nothing will do more to prolong the life of an engine.

A Clean Engine

A clean engine is as much a psychological factor in reliable performance as a mechanical one. The owner who keeps the exterior clean is more likely to care about the interior, and maintenance is less onerous. Nothing will put you off an overdue oil change more than the thought of having to crawl around a soot-blackened, dirty, greasy hunk of paint-chipped cast iron, with diesel fuel, old oil, and smelly bilge water slopping around in the engine drip pan.

Chapter 4

Troubleshooting, Part One: Failure to Start

It is useful to distinguish two differing situations when dealing with an engine that will not start. The first is a failure to crank—the engine will not turn over at all. The second is a failure to fire—the engine turns over but does not run.

SECTION ONE: FAILURE TO CRANK

When an engine will not crank at all, the problem is almost always electrical. Occasionally it is the result of water in the cylinders, or a complete seizure of the engine or transmission. Before you check the electrical system, see if you can turn the engine over by hand with the hand crank (if fitted), or by placing a suitably-sized wrench on the crankshaft-pulley nut. Turn the engine in its normal direction of rotation or you may accidentally undo this nut. If the engine has a manual transmission, you can also put it in gear and turn it with a pipewrench on the propeller shaft, but first wrap a rag around the shaft to avoid scarring the shaft.

If the engine is locked up solidly, it has probably seized and will need professional attention. If it turns over a little and then locks up, or turns with extreme difficulty, water may have siphoned into the cylinders

through the water-cooled exhaust (see Chapter 9 for an explanation of this, and how to avoid it). Check the oil level with the dipstick—if it's high, water in the engine is likely. (Note that the cylinders may have water in them, even with no rise in the oil level.)

Water in the Engine

Water, especially salt water, that remains for any length of time will do expensive damage to bearing and cylinder surfaces, requiring a complete engine overhaul. If you discover the water in time, you can ease it out of the exhaust and the engine will continue to operate.

Close the throttle so that the engine will not start. If the engine has decompression levers and a hand crank, simply turn it over several times. Otherwise, flick the starter motor on and off, or use a wrench on the crankshaft pulley nut to turn the engine over bit by bit, turning in the normal direction of rotation and pausing between each movement. Take it slowly—if you rush the process, you may damage piston rings and connecting rods.

Once the engine has turned through two complete revolutions, it should be free of water. Spin it a couple of times *without starting it*. Now check the crankcase

Troubleshooting Chart 4-1.
Engine Fails to Crank.

Note: Before jumping out solenoid terminals, completely vent the engine compartment. Try turning the engine over by placing a wrench on the crankshaft pulley nut. If the engine won't turn this is the problem—there may be water in the cylinders or it may be seized up.

Turn on some lights and try to crank the engine. Do the lights *go out?* **NO** ⬇	**YES** ▶ If the solenoid makes a rapid "clicking" the battery is probably dead—replace it. If the solenoid makes one loud click the starter motor is probably jammed or shorted—free it up or replace it as necessary.
When cranking, do the lights *dim?* **NO** ⬇	**YES** ▶ Check for voltage drop from poor connections or undersized cables. Try cranking for a few seconds, then feel all connections and cables in the circuits—battery and solenoid terminals, and battery ground attachment on engine block. If any are warm to the touch, they need cleaning. If this fails to show a problem set VOM to 12 volts DC. Place the probes according to Figure 4-10, then try to crank the engine. If there is no evidence of voltage drop, check for a jammed or shorted starter.
Is the ignition switch circuit faulty? **NO** ⬇ **TEST:** With a jumper wire or screwdriver blade, bridge the battery and ignition switch terminals on the solenoid. If the starter motor now cranks, the ignition circuit is defective.	**YES** ▶ Replace ignition switch or its wiring as needed.
Is the solenoid defective? **NO** ⬇ **TEST:** Use a screwdriver blade to jump out the two heavy-duty terminals on the solenoid. (This procedure is tricky: Observe all precautions outlined in accompanying text.) If the starter now spins, the solenoid is defective.	**YES** ▶ Replace the solenoid.
Is there full battery voltage at starter motor when cranking? **NO** ⬇ **TEST:** Check with VOM between the starter positive terminal and the engine block when cranking.	**YES** ▶ The starter is open-circuited and needs replacing. First check its brushes for excessive wear or sticking in their brush holders.
The battery isolation switch is probably turned "OFF"!	

for water in the oil. If any is present, change the oil and filter. Start the engine and run it for a few minutes to warm it, shut it down, and *change the oil and filter again*. Now give it a good run to drive out any remaining moisture. After 25 hours of normal operation, or at the first sign of any more water in the oil, *change the oil and filter for a third time*.

Put appropriate siphon breaks in the cooling and exhaust system (see Chapter 9).

Starter Motor Circuits

To troubleshoot electrical problems you first need to understand the starting circuit.

The heavy amperage draw of a starter motor requires heavy supply cables. Ignition switches frequently are located some distance from the ship's batteries and the engine. To avoid running the heavy cables to the ignition switch, a remotely operated switch—a *solenoid*—is placed in the starter motor circuit. Light-gauge wires run from this to the ignition switch (see Figure 4-1).

A solenoid contains a plunger and an electromagnet. When the ignition switch is turned on, it energizes the magnet, which pulls down the plunger, closes a couple of heavy-duty contacts, and completes the circuit to the starter motor (see Figure 4-2).

Some starting circuits utilize a *neutral start switch*, or solenoid, which prevents the engine from being cranked when it is in gear. Some circuits have a second solenoid. This energizes the first solenoid, which closes the starter motor points.

The ground (negative) side of a starter motor circuit almost always runs through the engine block to a ground strap connected directly to the battery's negative terminal. On occasion, however, you will find an insulated-ground starter motor, in which case the motor is electrically isolated from the engine block and has a separate ground cable to the battery. Although rarely done, this is the recommended practice in marine use because it helps in eliminating corrosion from stray current.

Inertia and Pre-engaged Starters

Starter motors are of two basic kinds: *inertia* and *pre-engaged*. The solenoid for an inertia motor is mounted independently at a convenient location. The starter motor drive gear that turns over the engine—the *pinion*—is keyed to a helical groove on the motor's drive shaft. When the solenoid is energized the motor spins. Inertia in the gear causes it to spin out along the helical groove and into contact with the gear on the circumference of the engine's flywheel. When the two engage, the engine turns over (see Figure 4-3).

Sometimes an inertia starter's pinion sticks to its shaft and won't engage the flywheel. In this case the starter motor will *whir* loudly without turning over the

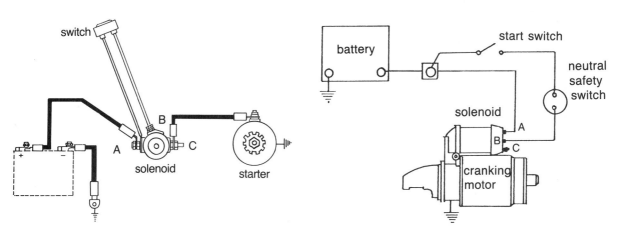

Figure 4-1. *Starting circuits. To bypass the switch, connect a jumper from A to B. To bypass the switch and the solenoid, connect a heavy-duty jumper (a screwdriver, for example) from A to C. (Courtesy PCM)*

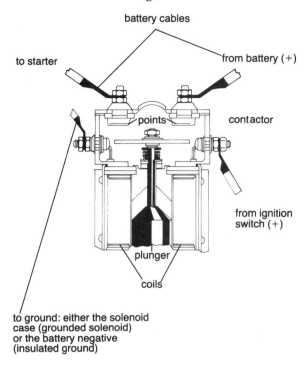

battery cables

to starter

from battery (+)

points

contactor

from ignition switch (+)

plunger

coils

to ground: either the solenoid case (grounded solenoid) or the battery negative (insulated ground)

engine. A smart tap on its case while it is spinning may free the pinion. (Don't do this too often—it's hard on the gear teeth. Note that if the starter continues to whir without working, the teeth may be stripped off its pinion or the flywheel ring gear—see page 50.) At other times, the pinion may jam in the flywheel and not disengage. If this happens, take off a cover on the rear housing of the starter motor and turn the squared-off end of the shaft back and forth with a wrench until you free the pinion.

The solenoid of a pre-engaged starter is *always* mounted on the starter itself (a surefire indicator—see Figure 4-4). When the solenoid is energized, the electromagnet pulls a lever, which pushes the starter-motor pinion into engagement with the engine's flywheel. The main solenoid points now close, allowing current to flow to the motor, which spins at full speed and cranks the engine. In this way, pre-engaged starters mesh the drive gear with the flywheel *before* the motor spins, greatly reducing wear on the drive gear and flywheel ring gear. (Note that if a pre-engaged starter whirs without cranking the engine, the teeth are almost certainly stripped off its pinion or the flywheel ring gear.)

Battery Testing

Note: never crank, or try to crank, a starter motor for more than 30 seconds continuously when you perform any of the following tests. Because starter motors are designed only for brief and infrequent use, they generally have no fans or cooling devices. Continuous cranking will burn them up.

Turn on a couple of lights and try to crank the engine. The lights should dim but still stay lit. If the lights remain unchanged, no current is flowing to the starter; if the lights go out, the battery is dead or the starter is shorted. Check the battery first.

The most effective way to check the battery's state of charge is with a *hydrometer*. (Note: this cannot be done with sealed batteries.) A hydrometer has a float, weighted on its bottom end, with a scale on its side. In pure water, the float reads 1.00 on the scale—known as the *specific gravity* of water.

A fully charged battery contains a solution of sulfuric acid, which is denser than water; therefore, a hydrometer floats higher in this liquid than it does in pure water. As the battery discharges, the sulfuric

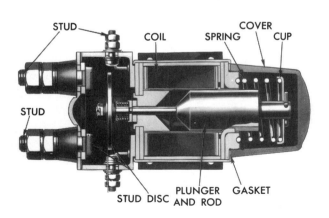

STUD

COIL

SPRING

COVER

CUP

STUD

STUD   DISC

PLUNGER AND ROD

GASKET

Figure 4-2. *Solenoid operation. When you turn the ignition switch to the start position, a small current passing through the solenoid's electromagnet pushes the contactor into the points, completing the circuit from the battery to the starter motor. (Courtesy Detroit Diesel Corp.)*

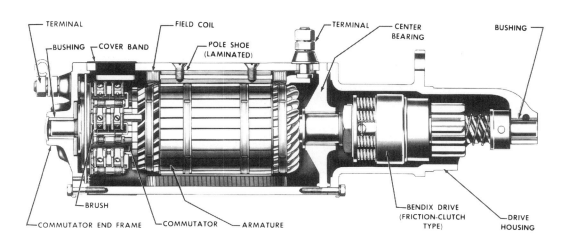

Figure 4-3. *Inertia-type starter motor. (Courtesy Detroit Diesel Corp.)*

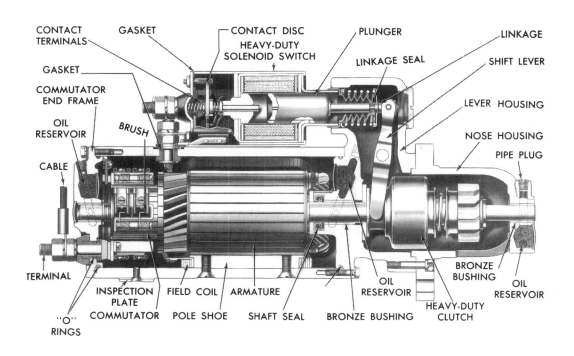

Figure 4-4. *Pre-engaged starter motor with solenoid mounted on it. (Courtesy Detroit Diesel Corp.)*

acid turns to water and the hydrometer sinks lower. A sample of the acid solution is withdrawn from each battery cell in turn. A fully charged battery will give a float reading of around 1.265 at 80°F (26.7°C), a half-discharged battery around 1.190, and a fully discharged battery around 1.120. These readings must be adjusted for changes in temperature (see Figure 4-5 and Figure 4-6).

Each battery cell must be tested individually. A battery is only as good as its weakest cell; it may have five good cells and one dead one. Once a battery refuses to hold a charge, you must replace it. (For an extensive discussion of batteries and boat electrics see my *Boatowner's Mechanical and Electrical Manual*.)

Circuit Tests

Quick Circuit Tests*. Note: some of these tests will create sparks. Be sure to vent the engine room properly, especially in the presence of gasoline engines.*

Assuming the battery is charged, a couple of quick tests will isolate problems in the starting circuit. The solenoid has two heavy-duty terminals: one holding the battery's positive cable, the other with a second

cable (or short strap in the case of a pre-engaged starter) running to the starter itself. There will also be one or two small terminals. If only one of the small terminals has a wire attached to it, this is the one you need. If both terminals have wires attached, one is a ground wire; you need *the other one*. (It goes to the ignition switch.) If in doubt and if you have an ohmmeter on board and know how to use it (see *Boatowner's Mechanical and Electrical Manual*), turn off the ignition switch and the battery isolation switch and test from both terminals on the solenoid to ground with the meter on the R × 1 scale. The ignition switch wire should give a small reading; the ground wire will read zero ohms.

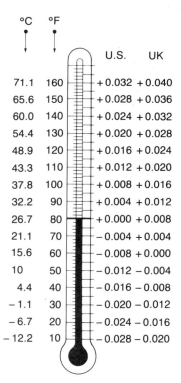

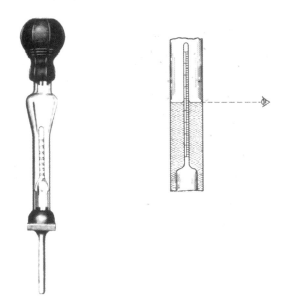

Figure 4-5. *A hydrometer for testing batteries. The correct method of reading it is shown at the right. Your eye should be level with the surface of the liquid. Disregard the curvature of the liquid against the glass parts.*

Figure 4-6. *Temperature corrections for hydrometer readings in the United States and the United Kingdom. Baseline temperature is 80°F (26.7°C) in the U.S. and 60°F (15.6°C) in the UK. Given a hydrometer reading of 1.250 and an electrolyte temperature of 20°F, you would subtract .024 in the U.S. to get a corrected specific gravity of 1.226. A fully charged battery has a specific gravity of 1.265, so our example is about 25 percent discharged.*

Turn on the battery isolation switch. Bypass the ignition switch circuit (and the neutral switch if fitted) by connecting a jumper wire or screwdriver blade from the *battery terminal* on the solenoid to the *ignition switch terminal* (see Figure 4-7 and Figure 4-8). If the motor cranks, the ignition switch or its circuit is faulty. (Note that key-type switches often malfunction in a marine environment and are best replaced with a sturdy push-button switch.) If the solenoid clicks but nothing else happens, the starter motor is probably defective (or the battery is flat—check it again). If nothing at all happens, the solenoid isn't getting current (check the battery isolation switch) or the solenoid is defective.

If the starter motor failed to work, use the screw-

Figure 4-8. *Bypassing the ignition switch by jumping out the two smaller wires.*

Figure 4-7. *Typical pre-engaged starter motor and solenoid. Jumping across 1 and 2 bypasses the ignition switch; 1 and 3 bypasses the ignition switch and the solenoid. The starter should spin but not engage the engine's flywheel.*

Figure 4-9. *Bypassing the solenoid altogether by jumping out the two main cable terminals.*

driver to jump out the two *heavy-duty terminals* on the solenoid (see Figure 4-9). *Be warned: the full starting current will be flowing through the screwdriver blade, and considerable arcing is likely.* A big chunk may be melted out of the screwdriver. Do not touch the solenoid case or starter case with the screwdriver: this will create a dead short. Hold the screwdriver *firmly* to the terminals. If the starter now spins, the solenoid is defective. If the motor doesn't spin, *it* is probably faulty. If no arcing occurred, there is no juice to the solenoid.

You can crank an engine that has an inertia starter by simply jumping a defective solenoid. A pre-engaged starter, however, will merely spin without engaging the flywheel because the solenoid is needed to push the starter pinion into engagement with the flywheel. In this case, have someone hold the start switch in the "ON" position while you jump the solenoid terminals. If the problem is just in the main solenoid points (the most likely bet), this will crank the engine, but if the problem is in the solenoid coil, the pinion will still fail to engage the flywheel.

Voltage Drop Tests. The above tests will quickly and crudely determine whether there is juice to the starter motor, and whether or not it is functional. More insidious is the effect on cranking performance of poor connections and undersized cables—these result in a *voltage drop*, which robs a starter motor of power. Less power in the starter motor means sluggish cranking or a failure to crank, especially when an engine is cold.

You can make a quick test for voltage drop by cranking (or trying to crank) for a few seconds then feeling all the cables and cable connections in the cranking circuit: both battery terminals, solenoid terminals, starter motor terminal, and the battery ground attachment point on the engine block. If any of these are warm, it indicates a resistive connection that needs cleaning. If any cable is warm, it indicates it is not large enough to carry the load being placed upon it.

Proper marine installations require an isolation switch that completely removes the battery from the ship's circuits in the event of an emergency, such as an electrical fire. Isolation switches are generally located in the boat's main electrical panel, which is frequently some distance from the battery and the starter motor.

In this case, the only way to reduce power losses from voltage drop in the long cable runs is to use larger-than-normal wire sizes. Welding cable, available from welding-supply houses, is often a good choice, although it must be carefully routed and supported since its insulation is rather soft and easily cut. Marine-grade cable, with tougher insulation, is available through most marine mail-order catalogs, but is several times more expensive.

Accurate voltage drop tests can be made with a voltmeter set on an appropriate scale to read 12 (or 24) volts DC. Run the positive meter probe to the *positive* post of the battery (not the cable clamp), and the other probe to the *positive* terminal of the solenoid. Crank the engine. You should have NO voltage reading. (See Figure 4-10—switch the meter to its lowest DC volts scale to get the most accurate reading.) *Any voltage reading shows voltage drop in this part of the circuit.* Clean all connections and repeat the test.

Perform the same test, cranking the engine each time, across the two main solenoid terminals, from the solenoid output terminal to the starter motor hot terminal, from the starter motor case to the engine block, and from the engine block to the negative post of the battery (not the cable clamp). There should be no voltage readings at any time. This test frequently reveals poor ground connections.

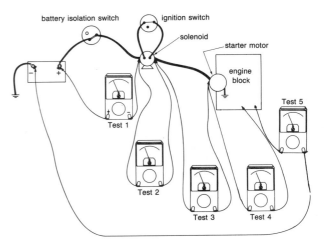

Figure 4-10. *Testing for voltage drop on starter motor circuits. Use a VOM set to 12 VDC; test while the engine is cranking to identify areas of excessive resistance.*

Motor and Solenoid Disassembly, Inspection, and Repair

Before removing the starter motor from an engine, isolate its hot lead, or better still, disconnect it from the battery. The starter motor is held to the engine by two or three nuts or bolts. If it sticks in the flywheel housing, a smart tap will jar it loose (but check first to see that you didn't miss a mounting bolt).

Inertia Starters. Remove the metal band from the rear of the motor case. Undo the locknut from the terminal stud in the rear housing. Do not let the stud turn: if necessary, grip it carefully with Vise-Grips (mole wrenches). Note the order of all the washers. These insulate the stud, and they *must* go back the same way.

Undo the two (or four) retaining screws in the rear housing. If they are tight, grip them with a pair of Vise-Grips from the side to break them loose (see Figure 4-11). Lift off the rear housing with care; it is attached to the motor case by two brush wires. Lift the springs off the relevant brushes and slide the brushes out of their holders to free the housing.

Pre-engaged Starters. To check the solenoid points on a pre-engaged starter, remove the battery cable, the screw or nut retaining the hot strap to the motor, and all other nuts on the stud(s) for the ignition circuit wiring. Undo the two retaining screws at the very back of the solenoid. The end housing (plastic) will pull off to expose the *contactor* and *points* (see Figure 4-12). (*Note: the spring in here that will probably fall out goes on the center of the contactor.*) Check the points and contactor (see Figure 4-13) for

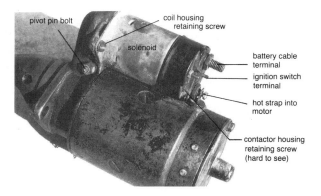

Figure 4-12. *Disassembling a pre-engaged starter.*

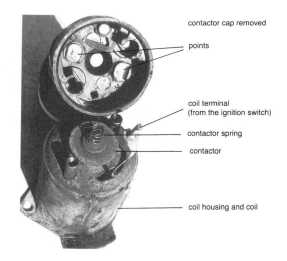

Figure 4-13. *Solenoid disassembly.*

Figure 4-11. *Removing stubborn bolts from the end plate.*

Figure 4-14. *Coil unit removed.*

note brush
and brush lead
retaining screw

insulated brush

brush grounded
to frame

commutator
segments

Figure 4-15. *End housing removed.*

pitting and burning and clean or replace as necessary. A badly damaged contactor frequently can be detached from its solenoid plunger and reversed to provide a new contact surface. In a similar way, the main points can sometimes be reversed in the solenoid end housing, but take care when removing them because the Bakelite housing is brittle.

To remove the *solenoid coil* (electromagnet), undo the two screws at the flywheel end of the solenoid and turn the coil housing through 90 degrees. The coil should now pull straight off the piston (see Figure 4-14). The *piston and fork* assembly generally can only be removed by separating the starter motor case from its front housing (see below).

If you find that you have to undo the *pivot pin* bolt on the solenoid fork, mark its head so that you can put it back in the same position. Often, turning this pin adjusts how far the pinion is thrown when it engages the flywheel.

Remove the two starter-motor retaining screws from the rear housing and lift off the rear cover; it comes straight off and contains no brushes. This exposes the *brushes* and *commutator* (the segmented copper bars on the rear end of the motor shaft—see Figure 4-15).

All Starters. Now you can pull off the motor case to expose the motor shaft (see Figure 4-16 and Figure 4-17). If the end cover was held with four short screws as opposed to two long ones, you'll find four more screws holding the motor case at the other end. Note that both end housings will probably have small lugs so that they can be refitted to the motor case in only one position. Inside the case will be the *field*

windings—densely packed copper coils enclosed in an insulating material. On the motor shaft will be the *armature*—another series of densely packed copper coils. Various tests can be performed on these coils with an ohmmeter, but these are beyond the scope of this book (see *Boatowner's Mechanical and Electrical Manual*).

Inspect the commutator and brushes for wear or signs of burning (pitting on segments of the commutator—see Figure 4-18). You can clean a commutator by pulling a strip of fine sandpaper (400-600 grit wet and dry) lightly back and forth until all the segments are uniformly shiny (see Figure 4-19). Cut back the insulation between each segment of the commutator to just below the level of the copper segments by drawing a knife or sharp screwdriver across each strip of insulation. Take care not to scratch the copper or burr its edges (see Figure 4-20). Use a triangular file to bevel the edges of the copper bars. Always renew the brushes at this time.

Brushes in any case need replacing if their length is much less than their width. Some brush leads are soldered in place; others are retained with screws. In replacing soldered brushes, cut the old leads, *leaving a tail long enough to solder to.* Tin this tail and the end of the new brush wires and solder them together. The new brushes *must slide in and out of their holders without binding*—if necessary, very gently sandpaper them down to make this possible.

The brushes attached to the field windings are insulated; those to the end housing or motor case are uninsulated. When replacing the end housing of an inertia starter, hold the brushes back in their holders and jam them in place by lodging the brush springs against the *sides* of the brushes. Once the plate is on, slip the springs into position with a small screwdriver.

commutator

armature

Figure 4-16. *Starter motor armature and commutator.*

case with field coils
and brushes

solenoid plunger

pinion

clutch

end housing
with bushing

solenoid plunger
forks

solenoid coil with
contactor and spring

armature and
commutator

solenoid contactor
housing

Figure 4-17. *A disassembled pre-engaged starter motor.*

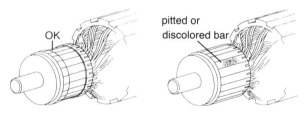

OK

pitted or
discolored bar

Figure 4-18. *Maintenance procedures for starter motors. Worn and grooved commutator is OK, as long as the ring is shiny. A pitted or dark bar results from an open or short circuit in the armature winding.*

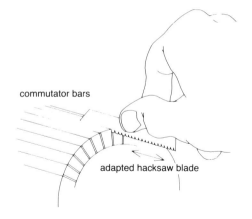

commutator bars

adapted hacksaw blade

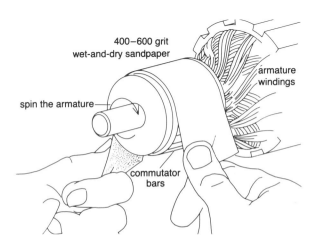

400–600 grit
wet-and-dry sandpaper

armature
windings

spin the armature

commutator
bars

Figure 4-19. *Polishing a commutator.*

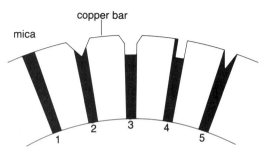

copper bar

mica

1   2   3   4   5

Figure 4-20. *Cutting back the insulation on a commutator. Modify a hacksaw blade as shown and run it between the commutator bars. When you start, the mica insulation should be flush (1). Cutting back as in 2 is good; 3 is better. Avoid cutting back as in 4 or 5.*

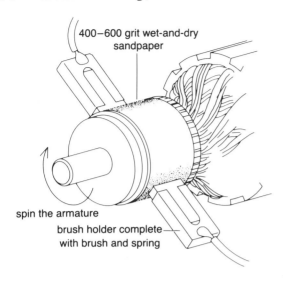

Figure 4-21A. *Bedding in new brushes.*

The commutator ends of new brushes will need bedding in. Wrap fine sandpaper (400 or 600 grit wet and dry) around the commutator under the brushes with the sanding surface facing out (see Figure 4-21). Spin the armature until the brushes are bedded to the commutator; they should be almost shiny over their whole surface. Remove the sandpaper and blow out the carbon dust.

The Pinion. Given the intermittent use of boat engines and the hostile marine environment, the pinion (or drive gear, often known as a *Bendix*) can be especially troublesome. Pinions are particularly prone to rusting, especially where salt water in the bilges has contacted the flywheel and been thrown all around the flywheel housing.

Inertia starters are more likely to give trouble than pre-engaged starters. Dirt or rust in the helical grooves on the shaft will prevent the pinion from moving freely in and out of engagement with the flywheel.

A small resistance spring in the unit keeps the pinion away from the flywheel when the engine is running. If this spring rusts and breaks, the gear will keep vibrating up against the flywheel with a distinctive rattle.

In manual-transmission automotive use, the shaft and helical grooves are not oiled or greased because the lubricant picks up dust from the clutch plates and causes the pinion to stick. Marine engines, however, don't have clutch plates in the flywheel housing so the shaft should be greased lightly to prevent rusting.

The pinion assembly is usually retained by a spring clip (a snap ring or circlip) on the end of the armature shaft—occasionally by a *reverse-threaded* nut and cotter pin (split pin). An inertia starter has a powerful *buffer* spring that must be compressed before the clip can be removed. Special tools are available for this, but a couple of small C-clamps or two pairs of adroitly handled Vise-Grips will do the trick.

Flywheel Ring Gear

It sometimes happens that the gear teeth get stripped off a section off the flywheel ring gear. Whenever the engine stops with this section of the flywheel lined up with the starter motor pinion, the starter will simply spin without cranking the engine.

If there are only a few teeth missing from the ring gear, as a temporary expedient you can turn the engine over fractionally by hand, either with a hand crank (if fitted) or else with a wrench on the crankshaft pulley nut (turn in the normal direction of rotation). This will line up some good teeth on the ring gear with the starter motor pinion, and should enable the engine to be started. However, every time the bad teeth come past the starter pinion, the starter motor

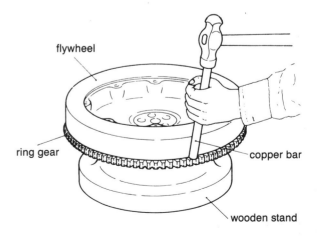

Figure 4-21B. *Renewing a flywheel ring gear. (Courtesy Yanmar Diesel Engines, U.S.A.)*

will momentarily race and then its pinion will crash into the good teeth on the ring gear as the flywheel's momentum brings them around. This will steadily tear up more and more teeth, so you need to attend to the problem as soon as possible.

To change a starter pinion, see the preceding sections on starter motor disassembly. The ring gear is an *interference* fit on its flywheel. To remove it, you must unbolt the flywheel from its crankshaft, and to do this you have to take the transmission out of the boat. You then place the flywheel on a stand and heat the ring gear evenly with a propane torch until it has expanded enough to allow you to drive it off with a hammer and copper bar (or something similar—see Figure 4-21B). To fit a new ring gear, pack the flywheel in ice while evenly heating the new gear, which should them just about drop into place. Position the new gear in exactly the same spot as the old one.

SECTION TWO: FAILURE TO FIRE

When I was working on oil platforms in the Gulf of Mexico, the most common emergency call I received went something like this: "Such-and-such engine won't start. It was working fine the last time I used it, but now it just won't run." I usually asked three questions before calling up a boat or helicopter to go out and investigate:

1. "Have you checked to see if it has any fuel?"
2. "Have you checked to see if the fuel filter is plugged up?"
3. "Have you left any life in the battery?"

"Oh, sure," was the answer, and probably half the time I found the engine out of fuel, the filter plugged, the battery dead, or a combination of these. Fuel can be added easily, and a clean filter can be fitted, but you can't get around a dead battery.

If an engine will not start as usual, you need to stop cranking and start thinking. Those extra couple of cranks in the hope of some miracle happening very often guarantee that the engine cannot be started at all. Most starting problems are simple ones that can be solved with a little thought—and frequently in a whole lot less time than it will take you to recharge a dead battery.

A diesel engine is a thoroughly logical piece of equipment. If its airflow is unobstructed, the air is being compressed to ignition temperatures, and the fuel injection is correctly metered and timed, it more-or-less has to fire. Troubleshooting an engine that won't start, therefore, boils down to finding the simplest possible procedure to establish which of these preconditions for ignition is missing.

1. An Unobstructed Airflow

A failure to reach ignition temperatures or problems with the fuel supply, are the most likely causes for starting failures. The airflow is the easiest to check, however, so let's investigate this first.

Does the engine have air? A stupid question, you may think, but certain engines (notably Detroit Diesels) have an emergency shutdown device, a flap that completely closes off the air inlet to the engine and guarantees that no ignition will take place. Once the flap is activated, even if the remote operating lever is returned to its normal position the flap will remain closed until manually reset *at the engine*. I have on several occasions flown to a platform to investigate an engine that would not start only to find the air flap closed. (*Note that stopping an engine by closing the air flap will soon damage the supercharger's air seals and should be done only in an emergency, such as engine run-away—more on this later.*)

If the engine doesn't have an air flap, what about the air filter? It may be plugged, especially if the engine has been operated in a dusty environment. It may have a plastic bag stuck in it, or even a dead bird (which I found on one occasion). If the boat has been laid up all winter, a bird's nest may be in there.

Does the engine have a turbocharger? Poor oil change procedures, or operating behavior (particularly racing the engine on start-up and just before shut-down—more on this later) may have caused the shaft to freeze in its bearings. Remove the inlet ducting and use a finger to see if the compressor wheel turns freely.

The other side of the airflow equation is the ability to vent the exhaust overboard. Starting problems, particularly on Detroit Diesels, may sometimes be the result of excessive back pressure in the exhaust. The most obvious cause would be a closed seacock. Other possibilities are excessive carbon build-up in the

Troubleshooting Chart 4-2.

Engine Cranks but Won't Fire.

Note: See Chart 4-1 if engine won't crank.

Is the air flow obstructed? Check any air flaps, air filter, and exhaust seacock for blockage or closure. **NO**	**YES**	Open air flap; replace air filter element; open exhaust seacock.
Is the engine cranking slowly? Note: Stop cranking and save the battery! **NO**	**YES**	Check for: low battery, voltage drop, improper oil viscosity. Try the five methods for boosting speed listed in the text under "Cranking Speed". If slow cranking is due to cold, see below. If these fail, recharge the batteries.
Is the engine too cold? **NO** Check cold-start devices. If glow plugs and manifold heaters are working, the cylinder head will be noticeably warmer. Plugs can be tested by using a multimeter, or unscrewing the plug and holding it against a good ground. See text under "Cold Start Devices" for details.	**YES**	Replace faulty glow plugs or manifold heaters; warm the engine, inlet manifold, fuel lines, and battery using a hair dryer, light bulb or kerosene lantern. Raise temperature slowly and evenly—concentrated heat can crack the engine castings.
Is the compression inadequate to achieve ignition temperature? **NO** (a) Suspect inadequate cylinder lubrication or piston blow-by. . . . (b) Suspect valve blow-by or blown cylinder head gasket. . . . (c) On Detroit Diesels, is the blower defective?	**YES**	**FIX:** On engines with custom-fitted oil cups on the inlet manifold, fill cups with oil and then crank engine. On others remove air filter and squirt oil into the inlet manifold as close to the cylinders as possible *while* cranking. See "Compression" in text. Check for incorrect valve clearance or head gasket blown. **FIX:** If valves are poorly seated, a top-end overhaul is needed. **FIX:** Consult a Detroit Diesel manual.
Is the fuel level too low? Check the fuel level in the tank. **NO**	**YES**	Add fuel. It will probably also be necessary to bleed the fuel system (see page 58).
Is the fuel delivery to the engine obstructed? **NO** Check to see that no kill devices are in operation; all fuel valves are open; no fuel filters are plugged; the remote throttle is actually advancing the throttle lever on the engine; and any fuel solenoid valve is functioning.	**YES**	If stop or kill control has been pulled out, push it in. Check power supply to and operation of fuel shutdown solenoid valve by connecting it directly to the battery with a jumper wire. If see-through fuel filters are plugged, change filters. Open the throttle wide. Push in the fuel knob on a Detroit Diesel hydraulic governor.
Is the fuel delivery to the injectors obstructed? **NO** **TEST:** Open throttle wide, loosen an injector nut and crank the engine. (But not on Detroit Diesels—see text)	**YES**	If no fuel spurts out, check primary, secondary, and lift pump filters and bleed the system (page 58). Check fuel lift pump for diaphragm failure. If fuel still does not flow, go back and check system for fuel level, blockages, and air leaks. Only after all else has been eliminated, suspect injection pump failure.
Note: If fluid spurts out when conducting previous test, make sure it is fuel, not water.		
If you have exhausted these tests, you can suspect incorrect timing, a worn fuel-injection pump, worn or damaged injectors.	**YES**	**FIX:** Replace pump or injectors. Timing problems indicate a serious mechanical failure; correction requires a specialist.

exhaust piping or in the turbocharger. In cold weather, there could be frozen water in a water-lift-type muffler, which has the same effect as a closed seacock.

The point is, take nothing for granted. A problem with the airflow is unlikely, but you need only a few minutes to check it out.

2. Ignition Temperatures

If you find no obstruction in the airflow, perhaps the air charge is not being adequately compressed to achieve ignition temperatures. Although numerous variables may be at work here, you must attempt to isolate them to identify problems.

Cold-Start Devices

The colder the ambient air, the lower its temperature when compressed, and the harder it is to get it up to ignition temperatures. As if this were not problem enough, cold thickens engine oil, which makes the engine crank sluggishly. Slower cranking gives the air in the cylinders more time to dissipate heat to cold engine surfaces and more time to escape past poorly seated valves and piston rings. A battery that puts out 100% at 80°F (27°C) will put out 65% at 32°F (0°C), and only 40% at 0°F (−18°C).

Cold is a major obstacle to reliable engine starting; therefore, most engines incorporate some form of a cold-start device to boost the temperature of the air charge during initial cranking. The most common device is a glow plug—a small heater installed in a pre-combustion chamber. Glow plugs are run off the engine-cranking battery, becoming red hot when activated.

Direct-combustion engines generally cannot use glow plugs because the combustion chamber doesn't have enough room for them. In this case the incoming air charge generally passes over some kind of a heater in the air inlet manifold—perhaps a heating element, or a *flame primer* (a device that ignites a diesel spray in the inlet manifold, thus warming the entering air— see Figure 4-22). Other engines (particularly Detroit Diesels) use a carefully metered shot of starting fluid to trigger the initial combustion process. (*But note that starting fluid should NOT be used in most*

instances—see below.) Additional heat is sometimes provided by heaters in the oil sump, in fuel filters, and around fuel lines.

If glow plugs and manifold heaters are working, the cylinder head or manifold will be noticeably warmer near the individual heating devices. If they are not working, first check the wiring in the circuit. There is frequently a solenoid activated by the ignition switch, or a pre-heat switch, in the same way that a starter motor solenoid is activated by the ignition switch. The same kinds of circuit tests can be made to this circuit as to a starting circuit (see Failure to Crank, above).

Glow plugs can be checked further by unscrewing them, holding them against a good ground (the engine block), and turning them on. They should glow red hot. You also can use an ammeter to test the power drain (5-6 amps per plug), or an ohmmeter to test the resistance (around 1.5 ohms per plug).

Test the amp draw by placing a suitable DC ammeter in the power supply line between the main hot wire and each glow plug *in turn* (not in the main harness itself since this may carry up to 40 amps on a six-cylinder engine). Test resistances by disconnecting the hot wire from each glow plug and testing from the hot terminal on the plug to a ground, using the most sensitive ohms scale on the meter (see Figure 4-23). Only a good ohmmeter will be accurate enough to distinguish between a functioning glow plug and a shorted plug.

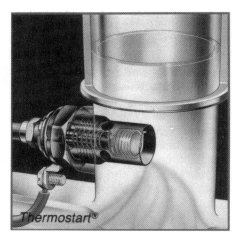

Figure 4-22. *A flame primer (Thermostart). (Courtesy Lucas CAV Ltd.)*

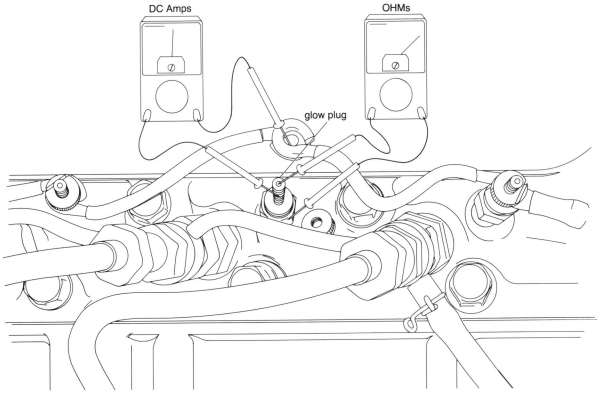

Figure 4-23. *Using a multimeter to test glow plugs.*

If a flame primer is not working, check its electrical connections and its fuel supply. If you remove the unit for further checks, you can see if the heating coil is working by jumping it from the battery, and you can check to see if the fuel is being discharged and atomized properly. *Never do these tests at the same time. If you operate the unit while it is removed from the engine you may receive nasty burns.*

Any safe means used to boost the temperature of the engine, battery, and inlet air will help with difficult starting. This includes using a hair dryer, light bulb, or kerosene lantern to warm fuel lines, filters, manifolds, and the incoming air; removing oil and water, warming them on the galley stove, and returning them; heating battery compartments with a light bulb; or removing the battery, putting it in a heated crew compartment, and returning it once it is warm.

A propane torch flame can also be used to boost temperatures. Gently play the flame over the inlet manifold and fuel lines, and across the air inlet when the engine is cranked so that it heats the incoming air. *A torch cannot be used in the presence of gasoline or propane vapors.* Do not play the torch flame over electrical harnesses, plastic fuel lines and fittings, or other combustibles.

Raise temperatures slowly and evenly, playing the heat source over a broad area. Concentrated heat may crack the engine's castings. Boiling water or very hot oil may do the same.

Compression

When an engine is operating, the lubrication system maintains a fine film of oil on the cylinder walls and the sides of the pistons. This oil plays an important part in maintaining the seal of the piston rings on the cylinder walls. After an engine is shut down, the oil slowly drains back to the crankcase. An engine that has been shut down for a long period may suffer a considerable amount of blow-by when you try to restart it because the lack of oil on the cylinder walls and piston rings reduces the seal, therefore compression.

An engine that grows harder to start over time is probably also losing compression, but this time due to poorly seated valves and piston-ring wear. (*Note that a Detroit Diesel exhibits the same symptoms if the blower is defective.*) The air in the cylinders is not compressed enough to produce ignition temperatures.

Short of a *top end* overhaul, there is not a lot that can be done with valve blow-by. Piston blow-by, however, can be cured temporarily by adding a little oil to the cylinders. The oil dribbles down and settles on the piston rings, sealing them against the cylinder walls.

If you have plenty of battery reserve, set the throttle wide open and crank for a few seconds. Then let the engine rest for a minute. Three things will be happening: the injected diesel will be dribbling down onto the piston rings; the initial heat of compression will be taking the chill off the cylinders; and the battery will be catching its breath. Try cranking again.

If this fails, or if little battery reserve remains, introduce a small amount of oil directly into the engine cylinders.

Adding Oil to Cylinders.

On engines with custom-fitted oil cups on the inlet manifold (for example, many older Sabbs—see Figure 4-24), fill the cups with oil then crank the engine. On others, remove the air filter and squirt oil into the inlet manifold as close to the cylinders as possible, while you crank the engine (see Figure 4-25). The oil will be sucked in when the engine cranks. Let the engine sit for a minute or two to allow the oil to settle on the piston rings. After starting, the engine will smoke abominably for a few seconds as it burns off the oil—this is OK. Put the air filter back in place as soon as the engine fires.

When you apply oil to the cylinders, use only a couple of squirts in each cylinder. Oil is incompressible; too much will cause damage to piston rings and connecting rods. Also, keep the oil can clear of turbocharger blades—a touch of the can will result in expensive damage.

Oil used in this fashion is often a magic—albeit a temporary—cure for poor starting, but the engine needs attention to *solve* the compression problem. To isolate the problem further, inject some oil into each cylinder in turn, then crank the engine with the throt-

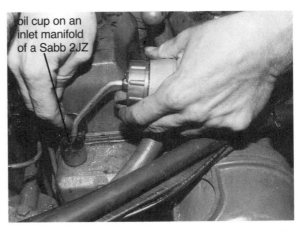

Figure 4-24. *Putting oil into an oil cup fitted to the air-inlet manifold of a Sabb 2JZ diesel.*

Figure 4-25. *Adding oil to the inlet manifold to increase combustion chamber pressure.*

tle closed so that it does not start. If you notice a marked improvement in compression on any cylinder, then you know this one is suffering blow-by around the piston and rings. If compression does not improve, suspect the valves.

A hand crank on the engine helps in performing these tests because you can slowly turn the crankshaft and rock each piston against compression. As each piston comes up to compression, you will feel the crank handle try to bounce back. Without the use of decompression levers, you should not be able to hand-crank a healthy engine through compression at slow speeds. If you can crank it through, you have considerable blow-by. You can frequently tell if the valves are the culprits by listening for a hiss of escaping air.

Compression Problems. Carbon build-up on valve stems, especially from prolonged low-load and low-temperature running (such as occurs when battery charging at anchor) will occasionally cause a valve to stick in its guide in the open position. This should not happen if you follow proper oil change procedures, but if tests indicate valve blow-by, remove the valve cover (rocker cover) and observe the valve stems for correct position. If work has recently been performed on the cylinder head or valves, it is also possible that one of the valve clearances has been improperly set (see Chapter 7), which can hold that valve in a permanently open position.

A loss of compression can also be caused by a blown, or leaking, cylinder-head gasket. The symptoms will resemble those of leaking valves, and because the head must be removed to sort out the valves, the gasket problem will become evident right away. Unless the head has been recently removed there is little reason to suspect the gasket.

Engine wear, especially in the piston pin and rod end bearings, eventually may increase the cylinder-head clearance to the point at which adequate compression cannot be achieved. Before this point is ever reached, the engine will give advance warning by *knocking* (see Chapter 5). A major engine rebuild will be needed.

Finally, it is possible for the operator of an engine fitted with decompression levers to leave these levers in the decompressed position, which ensures that the cylinder has no compression—silly, but worth checking before you take more drastic action. If an engine is shut down by using the decompression levers instead of closing the fuel rack on the injection pump, serious damage is likely to occur to the valves and push rods. If you suspect this might have happened, remove the valve cover and check for bent valve stems or push rods.

Cranking Speed

No diesel will start without a brisk cranking speed (at least 60 to 80 rpm: most small diesels will crank at

200 to 300 rpm). The engine, especially when it is cold, just will not attain sufficient compression temperature to ignite the injected diesel (see Figure 4-26). If the motor turns over sluggishly, *stop cranking and save the battery*: You will need all the energy it has.

First, check the battery's state of charge. If it is fully charged, check for voltage drop (which may be robbing the starter motor of power) between the battery and the starter motor (see page 46). Assuming a good battery and a properly functioning starting circuit, the techniques in the sections on cold-start devices and compression in this chapter will help generate that first vital power stroke. Sometimes the following tricks also will boost cranking speeds:

- If fitted with decompression levers and a hand crank, turn the engine over a few times by hand to break the grip of the cold oil on the bearings. Assist the starter motor by hand cranking until the engine gains momentum then knock down the decompression levers.
- Disconnect all belt-driven auxiliary equipment (refrigeration compressor, pumps, etc.) to reduce the starting load.
- Place a hand over the air inlet while cranking. Restricting the airflow will reduce compression and help the engine build up speed. Once the engine is cranking smartly, remove your hand—the motor should fire. *Never block the air inlet on an operating engine; the high suction pressures generated may damage the engine and your hand.*
- In a dire emergency, loosen one or two injectors in the cylinder head of an engine that doesn't have decompression levers so that

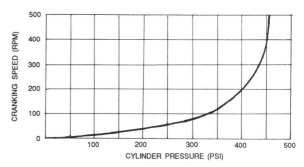

Figure 4-26. *Approximate relationship of engine cranking speed to cylinder pressure.*

these cylinders blow by, allowing the engine to pick up momentum. Once the engine starts, immediately tighten them. Be warned that on some engines—Volvos and Detroit Diesels, for example—this procedure may loosen the injector sleeves in the cylinder head as well as the injectors, which will let cooling water into the engine. Special tools will be needed to reseat the injector sleeves.

- An additional trick for *sailboats with manual transmissions* is to sail the boat hard in neutral with the propeller freewheeling, then start cranking and throw the transmission into forward. The additional momentum of the propeller may bump start the engine.

3. Fuel Problems

If an engine is cranking smartly, with sufficient compression to produce ignition temperatures, but still won't start, the culprit is almost certainly the fuel system. This has the potential for causing a considerable number of problems! Some are easy to check but others can only be guessed at.

Check the Obvious

Diesel engines are shut down by closing off the fuel supply. On some engines this occurs when the throttle is closed; others continue to idle at minimum throttle settings and a separate *stop* control is fitted to shut off the remaining fuel supply. Has the stop control inadvertently been left pulled out? Has an emergency shut-down device, such as the air flaps on Detriot Diesels, been inadvertently tripped? Is the throttle open to the position specified for starting by the engine manufacturer? A diesel will never start with the throttle closed. Trace the throttle cable from the throttle lever in the cockpit to the engine and make sure that it is actually advancing the throttle lever on the engine.

Note that a Detroit Diesel with a hydraulic governor must be cranked for a few seconds to generate enough oil pressure in the governor to open the fuel rack so that the engine can start. On the outside of the governor, you will see a rod with a knob on the end of it. If this is pushed in by hand and held while crank-

ing, it will by-pass the governor, opening the fuel rack.

Is there plenty of fuel in the tank? The fuel suction line is probably set an inch or two off the bottom of the tank; if the boat is heeling, air can be sucked in even when the fuel level appears to be adequate. Is the fuel valve (if fitted) open? If there is fuel and the valve is open, but no fuel is reaching the engine, you may have a small filter screen inside the tank on the suction line which has become plugged. If such a screen is fitted, *throw it away*—this is not the proper place to be filtering your fuel. This is why you have fuel filters.

Most engines have mechanical lift pumps, but a few have electrical pumps. With the latter you should hear a quiet clicking when the ignition is first turned on. If in doubt about an electric pump's operation, loosen the discharge line and see if fuel flows.

Solenoid Valve

Many newer engines have a solenoid operated fuel shutdown valve which is held in the closed position by a spring when the ignition is turned off. When the ignition is turned on, it energizes a magnet, which opens the valve. Any time the electrical supply to the solenoid is interrupted, the magnet is de-energized and the spring closes the valve. Any failure in the electrical circuit to the valve will automatically shut off the fuel supply to the engine.

Some solenoid fuel valves are built into the back of the fuel injection pump. You can identify these by a couple of wires coming off the pump close to the fuel inlet line. Others are mounted separately but close to the pump (see Figure 4-27). A rod coming from the back of the valve actuates a lever on the pump. You can check the operation of a solenoid valve by connecting it directly to the battery with a jumper wire. Take care to get the positive and negative leads the right way around. If the valve has only one wire, this is the positive lead; if two, one will run to ground and we want the other one.

Fuel Filters

The primary fuel filter should have a see-through bowl, which you should check for water and sedi-

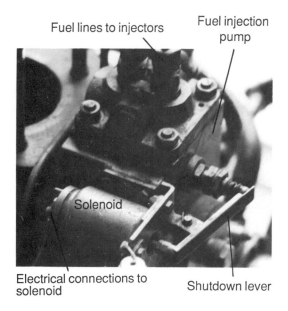

Fuel lines to injectors   Fuel injection pump

Solenoid

Electrical connections to solenoid

Shutdown lever

Figure 4-27. *Solenoid-operated fuel shut-down valve.*

Filter screen

Handle for manual operation

Figure 4-28. *Fuel filter in a lift pump.*

ment. If it is opaque, open the drain on its base and take a sample. On many fuel systems, this will let air into the system, which will then need *bleeding* (see below). It is not uncommon for a primary fuel filter to be completely plugged up. If this is the case, don't take chances—replace it along with the secondary filter; drain the tank or pump it down until all traces of contamination are removed. If filters repeatedly clog, it is likely that sediment has built up inside the tank to the level of the fuel pick-up. The tanks will need opening and flushing.

If there is no primary filter, or any sign of contamination making its way past the primary filter, you will generally find inside diaphragm-type lift pumps a screen that also needs checking. You can reach it by undoing the center bolt and removing the cover (see Figure 4-28).

Assuming the tank has fuel and the filters are clean, the next step is to find out if the fuel lines have air in them.

Bleeding (Purging) a Fuel System

Air trapped in the fuel system can bring most diesels to a halt, although the extent to which this is true var-

ies markedly from one engine type to another. Detroit Diesels can be purged simply by opening the throttle and cranking for long enough, whereas many older diesels with jerk-type fuel injection pumps can be completely disabled with even tiny amounts of air. Engines with distributor-type fuel injection pumps tend to fall somewhere between these two extremes. When air has to be purged from a fuel system by hand, the process is known as *bleeding*.

Tracing the Fuel Lines. With the exception of Detroit Diesels, typical fuel systems are shown in Figures 2-9 and 2-13. The fuel is drawn from the tank by a lift pump (sometimes called a feed pump) and passes through the primary filter. The lift pump pushes the fuel on at low pressure through the secondary filter to the injection pump. (On some engines, the lift pump is incorporated into the back of the fuel injection pump rather than being a separate item.) The injection pump meters the fuel and pumps exact amounts at precise times and at very high pressures, down the injection (*delivery*) lines to the injectors then into the cylinders. Any surplus fuel at the injectors returns to the secondary filter or tank via *leak-off*, or return, pipes.

The more cylinders an engine has, the greater the number of fuel lines. The various filters and lines can sometimes become a little confusing. Just remember that the secondary fuel filter is generally mounted on the engine close to the fuel injection pump, whereas the primary filter is generally mounted off the engine, or on the engine bed, closer to the fuel tank. The filters should have an arrow on them to indicate the direction of fuel flow; sometimes the ports will be marked "in" and "out."

There will be an injection line (delivery pipe) from the injection pump to every injector, but the leak-off pipes go from one injector to the next then down a common pipe to the secondary filter or fuel tank. This makes it easy to distinguish delivery pipes and leak-off pipes on most engines. A few engines, however (notably many Caterpillars), have *internal* fuel lines and injectors that are hidden by the valve cover. In this case, each delivery pipe runs from the fuel injection pump to a fitting on the side of the cylinder head, and from there all is hidden (see Figure 4-29). Detroit Diesels are as described on page 20.

At various points in the system you will find bleed nipples—normally on the filters and the injection pump. One should be located on the top of the secondary filter. On the base of the lift pump is usually a small handle, enabling it to be operated manually (see Figure 4-30). Pump this handle up and down. If it has

little or no stroke, the engine has stopped with the lift pump drive cam at or near the full stroke position. You will need to spin the engine a half turn or so to free the manual action. (See the section on lift pumps below for a more complete explanation of this.)

Engines that do not have an external lift pump generally have a manual pump attached to the injection pump, one of the filters, or at some other convenient point in the system (see Figure 4-31). Bleeding follows the same procedure as that used on a lift pump.

Bleeding the Low-pressure Lines. Open the bleed nipple on the secondary filter and operate the lift pump. If the filter has no bleed nipple, loosen the connection on the fuel line coming out of the filter (see Figure 4-32). Fuel should flow out of the bleed nipple or loosened connection *completely free of air bubbles.* If bubbles are present, you will have to operate the lift pump until they clear then close the nipple or tighten the connection. This should have purged the air from the suction lines all the way back to the tank, including both filters. If any of the fuel lines have a high spot, however, a bubble of air may remain at this point and

Figure 4-29A. *Internal fuel lines. This photo shows fuel lines inside a valve cover and the point at which they enter the engine. (Courtesy Caterpillar Tractor Co.)*

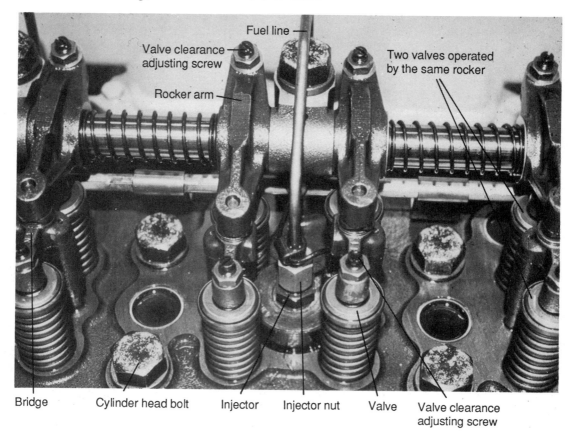

Valve clearance adjusting screw

Fuel line

Two valves operated by the same rocker

Rocker arm

Bridge   Cylinder head bolt   Injector   Injector nut   Valve   Valve clearance adjusting screw

Figure 4-29B. *This photo shows the point at which internal fuel lines connect to the injectors. This high-performance diesel has two inlet and two exhaust valves per cylinder. The rocker arm opens each set of valves via a bridge. (Courtesy Caterpillar Tractor Co.)*

be extremely hard to dislodge. Note that if the manual lever strokes but then fails to return to its original position, its internal spring is broken. This is the cause of the fuel supply problems—see page 64 for disassembly and repair procedures.

Take the trouble to catch or mop-up all vented fuel. Diesel will soften, and eventually destroy, most wire insulation and also the rubber feet on flexibly mounted engines.

The next step is to bleed the fuel injection pump. Somewhere on the pump body you will find one, or perhaps two, bleed nipples. (Some of the modern pumps are self-bleeding and have no nipples.) If the pump has more than one nipple, open the low one

first and operate the lift pump until fuel that's free of all air bubbles flows out (see Figure 4-33, Figure 4-34, and Figure 4-35). Close the nipple and repeat the procedure with the higher one. The injection pump is now bled.

Bleeding the High-pressure Lines. Bleeding the fuel lines from the injection pump to the injectors is the final step. To do this, set the governor control (throttle) *wide open* (this is essential) and crank the engine so that the injection pump can move the fuel up to the injectors. This should take no more than 30 seconds. In any event, a starter motor should not be cranked for longer than this at any one time because

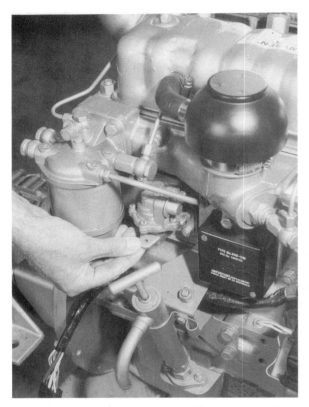

Figure 4-30. *Manual operation of a lift pump. (Courtesy Perkins Engines Ltd.)*

Figure 4-31. *Manual fuel pump on a filter.*

serious damage can result from internal overheating. If the engine has decompression levers and a hand crank, turn it over by hand to avoid running down the battery.

If the engine has no hand crank, you must properly bleed the system to the injection pump before attempting this last step. The battery is frequently already low because of earlier cranking attempts; therefore, pumping up the injectors one time, let alone having to come back to try again if rebleeding is necessary, will be touch and go.

When the fuel eventually reaches the injectors, provided the engine is not running, you can hear the moment of injection as a distinct creak. Familiarize yourself with this noise. If you can recognize the sound and if it is present when you make your first unsuccessful attempt to crank the engine, at least you know that fuel is reaching the engine and you don't

Figure 4-32. *Bleeding a secondary fuel filter. (Courtesy Perkins Engines Ltd.)*

Figure 4-33. *Bleeding the fuel inlet pipe to a distributor-type injection pump. (Courtesy Perkins Engines Ltd.)*

Figure 4-34. *Bleeding the lower nipple on a CAV DPA distributor-type fuel injection pump. (Courtesy Perkins Engines Ltd.)*

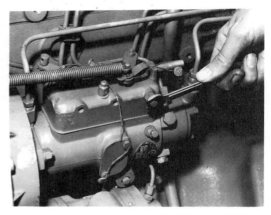

Figure 4-35. *Bleeding the upper nipple on a CAV DPA distributor-type fuel injection pump. (Courtesy Perkins Engines Ltd.)*

have to bleed it (unless you are injecting water!).

If the engine still will not fire, loosen one of the injector nuts (these hold the delivery lines to the injectors—see Figure 4-37 and Figure 4-38) and crank again. A tiny dribble of fuel, free of air, should spurt out of this connection at every injection stroke for this cylinder (every second engine revolution on a 4-cycle engine). If there is no fuel, or there is air in the fuel, the bleeding process has not been done adequately and needs repeating. *On engines with internal fuel lines, loosen the nuts on the delivery pipes where they enter the cylinder head, not the nuts on the injectors, so that none of the vented diesel gets into the engine oil* (see Figure 4-29A).

Do not overtighten injector nuts because this may collapse the fitting that seals the delivery pipe to the injector. Tighten the nuts to 15 pounds feet if you don't have the manufacturer's recommended setting. Any time you loosen an injector nut, you must check it for leaks when the engine is running. *Fuel leaks on engines with internal fuel lines will drain into the crankcase, diluting the engine oil and possibly leading to engine seizure* (see page 112 for how to check internal fuel lines).

Bleeding Detroit Diesels. Detroit Diesel (common rail) fuel systems are self-purging. As long as the tank has fuel in it, the suction line is free of breaks,

and the fuel pump works, the diesel flowing through the system will drive out any air. To check the fuel flow, undo the return line from the cylinder head to the tank then crank the engine—a steady flow should come out (not the little dribbles you see from jerk-pumps and distributor-pumps).

Persistent Air in the Fuel Supply

One of the more aggravating problems on many 4-cycle diesels can be persistent air in the fuel system. Air can come from poor connections, improperly seated filter housings (especially if the problem occurs after a filter change), and pinholes in fuel lines due to corrosion and vibration against bulkheads or the engine block. Since the only part of the fuel system under a suction pressure is that from the tank to the lift pump, this is the most likely problem area. Sometimes fuel tanks are deeper than the lifting capa-

bility of the pump; the pump may fail to raise the fuel when the tank is almost empty, or fail to raise enough fuel (called fuel starvation) at higher engine loadings.

If the primary filter has a see-through bowl, loosen the bleed nipple on the secondary filter, operate the hand pump on the lift pump, and watch the bowl. Air bubbles indicate a leak between the fuel tank and the primary filter, or in the filter gasket itself. No air: the leak is probably between the filter and the pump.

In the absence of a see-through bowl, locate the fuel line that runs from the lift pump to the secondary filter. Disconnect it *at the filter*. Place it in a jar of clean diesel and pump. If there is a leak on the suction side, bubbles will appear in the jar.

Any air source on the lift pump's discharge side should reveal itself as a fuel leak when the engine is running. When the engine is shut down, the fuel may suck in air as it siphons back to the tank. Next time the engine is cranked, it will probably start and then

Secondary fuel filter
bleed nipple   Fuel lines to injectors   Fuel pump bleed nipple

Figure 4-36. *Bleed points on a Volvo MD17C.*

Valve cover

Decompression lever (in decompressed position)

Injector hold-down nut

Injector leak-off pipe

Injector nut

Figure 4-37. *Injectors on a Volvo MD17C.*

Leak-off pipe

Injector hold-down bolts   Injector nut

Figure 4-38. *Location of injector nut.*

die. A similar problem can arise *when the leak-off pipe from the injectors is teed into a fitting on the secondary fuel filter with another (overflow) line running from here back to the tank* (this is a common arrangement). When the engine is shut down, fuel will sometimes siphon down the overflow line and cause air to be sucked into the system. In this situation, there is no external evidence of the air source, making for frustrating detective work. To cure the problem, either move the leak-off pipe so that it runs directly to the fuel tank, or add a length of flexible fuel line between

the injectors and the secondary filter. Loop this above the level of the filter and injectors and drill a small ($^1/_{16}$") hole in it at its highest point. This will act as a siphon break.

On rare occasions fuel may siphon back through a defective injector, the injection pump, and the lift pump, but a number of things have to be out of order for this to happen. Another rare source of air is the pump housing of some distributor-type fuel injection pumps with built-in vane-type lift pumps (see below). To test, run a line from the injection pump suction fitting straight into the fuel tank, and run one or more of the injector delivery pipes into a jar of clean diesel. Bleed the pump, purge the delivery pipes, open the throttle, and crank the engine. If air bubbles appear in the jar, the injection pump housings themselves are sucking air. Finally, if an engine tends to run for a good long while then suffers from fuel starvation or dies with air in the system, check the fuel tank vent to make sure it isn't plugged. A plugged vent causes a vacuum to form in the tank as the fuel is consumed.

Lift Pump (Feed Pump) Failure

Most small 4-cycle diesels found in boats use a diaphragm-type lift pump (see Figure 4-39). Larger engines and Detroit Diesels use a gear-driven pump (see Figures 4-40 and 4-41).

A diaphragm pump has a housing that contains a suction and a discharge valve, plus the diaphragm (see

Figure 4-39). A lever, which is moved up and down by a cam on the engine's camshaft or crankshaft, pushes the diaphragm in and out. This lever can also be activated manually, but if the engine is stopped in a position that leaves the diaphragm lever fully depressed, the manual lever will be ineffective until the engine is turned over far enough to move the cam out of contact with the lever.

Diaphragm pumps are nearly foolproof, but eventually the diaphragm will fail. When this happens, little or no fuel will be pumped out of the fuel system's bleed nipples when the lift pump is operated manually. Older pumps often have a drain hole in the base from which fuel will drip if the diaphragm has failed, but recent Coast Guard regulations have banned this for newer pumps.

A spare diaphragm, or better yet a complete pump unit, should be part of the spares kit on boats that cruise offshore. Diaphragms are accessible by undoing a number of screws (generally six) around the

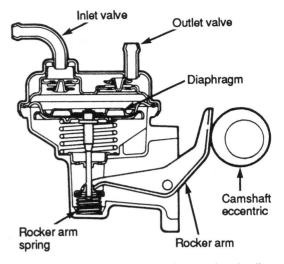

A mechanical fuel pump is mechanically actuated by a rocker arm or push rod without electrical assistance.

Figure 4-39. *A typical mechanical fuel pump. (Courtesy AC Spark Plug Division, General Motors Corp.)*

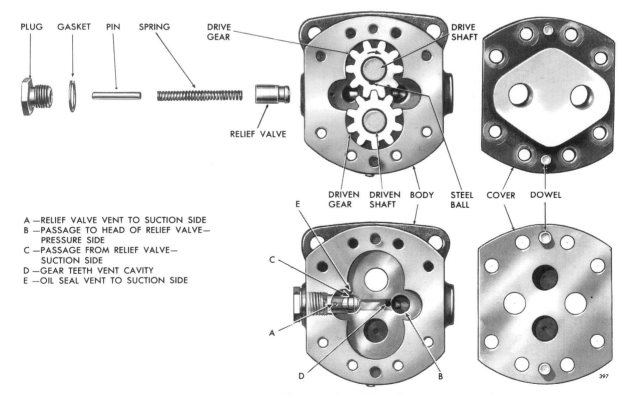

A —RELIEF VALVE VENT TO SUCTION SIDE
B —PASSAGE TO HEAD OF RELIEF VALVE—
    PRESSURE SIDE
C —PASSAGE FROM RELIEF VALVE—
    SUCTION SIDE
D —GEAR TEETH VENT CAVITY
E —OIL SEAL VENT TO SUCTION SIDE

Figure 4-40A. *This break-down of a gear-type lift pump shows how the various parts fit together. (Courtesy Detroit Diesel Corp.)*

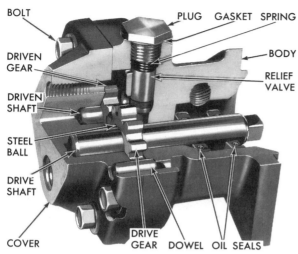

Figure 4-40B. *Cutaway of a gear-type lift pump shows how the parts relate to one another. (Courtesy Detroit Diesel Corp.)*

Figure 4-41. *This gear-type lift pump (transfer pump) is fitted to Caterpillar engines and performs the same function as a diaphragm lift pump. It doesn't have provision for hand pumping, so a hand pump is incorporated into one of the fuel filters. Unlike a diaphragm pump, a gear pump always puts out more fuel than the engine needs. The excess bleeds back to the inlet side of the pump via a pressure-relief valve built into the pump or installed elsewhere in the fuel system. (Courtesy Caterpillar Tractor Co.)*

body of the pump and lifting off the top half (see Figure 4-42). The method of attaching the diaphragm to its operating lever varies—it may be necessary to remove the whole pump from the engine (two nuts or bolts) and play with, or remove, the operating lever (see Figure 4-43).

Some newer engines have the lift pump built into the fuel injection pump. Two types are used:

1. On in-line jerk pumps there is likely to be a piston-style pump in which a plunger moves up and down via a cam on the same camshaft that operates the individual jerk pump plungers. These pumps generally incorporate an externally operated plunger for manual priming of the fuel system.
2. On distributor pumps, a rotary vane pump is driven off the central drive shaft. These pumps cannot be operated by hand; therefore, a separate manual pump is generally included in the system at some point, usually tacked onto one of the filters.

Very Cold Weather

The diesel fuel almost universally available in the United States is known as Number 2 diesel. At very low temperatures, it congeals enough to plug up fuel filters and lines. If you suspect this to be a problem, heating fuel lines and filters with a hair dryer or some other heat source may be all that's necessary to get things moving. If you anticipate prolonged extra-cold weather, thin Number 2 diesel with a special low-temperature additive, kerosene, or Number 1 diesel. *Please note, however, that all of these decrease the fuel's lubricating qualities; running some engines on straight Number 1 diesel, for example, can lead to engine seizure.*

Serious Fuel Supply Problems

If the air inlet and exhaust are unobstructed, compression is good, the tank has fuel, and the system is properly bled, it is time to feel nervous and check the bank balance. Not too many possibilities remain for a failure to start—basically a worn injection pump, worn or damaged injectors, or incorrect fuel injection timing.

There is just no reason for injection timing to go

diaphragm cover screw

diaphragm

manual operation lever

Figure 4-42. *Diaphragm on a lift pump.*

out unless the engine has been stripped down and incorrectly reassembled. Only some serious mechanical failure is going to throw out the timing, in which case you should have plenty of other indications of a major problem.

Worn or damaged injectors can lead to inadequate atomization of the fuel, to the extent that combustion fails to take place. Injectors are as precisely made as injection pumps and should only be disassembled as a last resort. Chapter 6 describes procedures for removing, cleaning, and checking the injectors.

If the pump is so badly worn that proper injection is no longer occurring, you can do nothing except have it rebuilt or exchange it for a new one. Changing injection pumps is covered under the section on engine timing in Chapter 7.

Let me emphasize that these problems will almost never occur in a well-maintained engine. Just about every other fault should be suspected before them.

Starting Fluid and WD 40

Unless specific provision for the use of starting fluid is made on an engine (e.g., some Detroit Diesels and

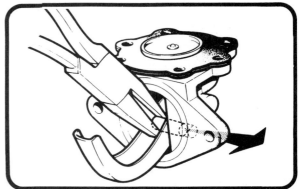

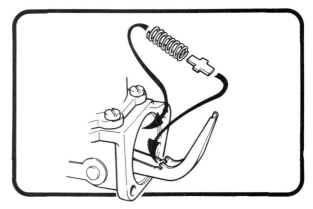

Figure 4-43. *Diaphragm replacement on a lift pump. (Courtesy Volvo Penta)*

Caterpillars), do not use it at all. It is sucked in with the air charge, and being extremely volatile will often ignite before the piston is at the top of its compression stroke. This can result in serious damage to pistons and connecting rods. For some reason, Detroit Diesels seem to tolerate starting fluid better than 4-cycle diesels.

If starting fluid must be used, do not spray it directly into the air inlet manifold. Rather, spray it onto a rag and hold this to the air intake. This will control the amount being drawn in. Adding a little diesel to the inlet manifold will also help the engine to pick up once the starting fluid fires.

Diesels will run on WD 40 (don't ask me why!) at far less risk of premature detonation than that imposed by starting fluid. In fact, if the battery is low but you need to do some extended cranking—such as when having problems purging a fuel system—you can open the throttle wide and spray a continuous stream of WD 40 into the air inlet while someone else cranks. The engine will fire and continue to run as long as you keep spraying. You can control the engine speed with the WD 40 until the fuel system is bled and the engine takes over. This is also an effective way to get a drowned engine running again, after the water has been driven out of the cylinders and the oil and filter changed.

Chapter 5

Troubleshooting, Part Two: Overheating, Smoke, Knocks, and Other Problems

Overheating

Overheating can be the result of a number of things, but the primary suspect is always a loss of flow in the raw-water circuit. For this reason, as well as to prevent following waves driving up the exhaust pipe, a water-cooled exhaust should exit fairly high in the stern. You will then be able to see at a glance whether the raw-water side of a cooling system is functioning. You should make an iron habit of checking the exhaust for proper water flow every time you start the engine.

Overheating on Start-up. Did you forget to open the raw-water seacock? If you did, since most raw-water pumps have a rubber impeller and since these cannot tolerate running dry, there is a good possibility that you have now stripped the vanes off the pump impeller.

To check a pump, first check the pump drive belt to make sure it isn't broken or slipping. If this is OK, remove the pump cover (normally six screws) and check the impeller. If the impeller is OK, before replacing the cover, turn the engine over and make

sure that the impeller is also turning; i.e., not slipping on its drive shaft. If the impeller is damaged, pull it out with needle-nose pliers or pry it out with a couple of screwdrivers (see Figure 5-1). Most impellers are a sliding fit on their shaft, but a few have locking screws (see Figure 5-2). Track down the missing vanes—they are likely to be found in the heat exchanger (if fitted). If the pump needs rebuilding for any reason, see page 95.

Faced with an irreparable failure of the raw-water pump, you can keep an engine running in an emergency by setting a bucket in the cockpit, disconnecting the engine's raw-water line from the raw-water pump, and running the line up into the bucket, using some extra hose. Once the line is purged of air, water will siphon from the bucket, through the raw-water side of the cooling circuit, and out of the exhaust. The rate of flow can be increased by raising the bucket into the rigging. Keep the bucket filled at all times, and keep a sharp eye out to make sure you don't lose your prime and that the engine doesn't overheat.

If the raw-water circuit is functioning as normal and the engine is freshwater cooled, check the level in the coolant recovery bottle (if fitted) or expansion

Troubleshooting Chart 5-1. Overheating on Start-up.		
Is water coming from the raw-water discharge? **NO**	**YES**	Check the coolant level in the fresh-water circuit (if fitted). Caution: do not remove header-tank pressure-cap when hot. If the level is low, refill and find the leak.
Is the raw-water seacock closed? **NO**	**YES**	Open and then check the raw-water overboard discharge. The raw-water pump may have failed from running dry.
Is the raw-water strainer plugged? **NO**	**YES**	Clean and then check raw-water discharge as above.
Has the raw-water pump failed? Inspect the drive belt and tension or replace as necessary. Make sure any clutch is operative. If the belt and clutch (if fitted) are OK, remove the pump cover and inspect the impeller vanes. Make sure the impeller turns when the engine turns. **NO**	**YES**	Tighten or replace the drive belt as necessary. Replace a damaged impeller. Track down any missing vanes. See also troubleshooting chart 5-2 on page 71.
Check for collapsed or kinked raw-water hoses, an obstruction over the raw-water inlet on the outside of the hull (break a below-the-waterline hose loose as close to the raw-water seacock as possible and see if there is a good flow into the boat), or a plugged raw-water injection nozzle into the exhaust. Is the water-lift silencer frozen?		

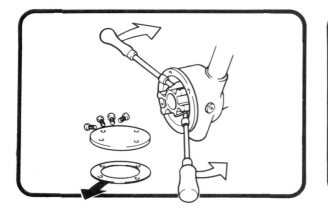 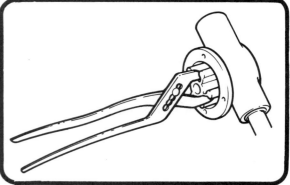

Figure 5-1. *Most flexible water-pump impellers can be pried or pulled out. (Courtesy Volvo Penta)*

tank. Warning: Never remove the cap when it's hot; serious burns may result. If the level is low, find out where the water is going. Possibilities are leaking hose connections, heat-exchanger cooling tubes that have corroded through (see "Water in the Crankcase" later in this chapter), or a blown head gasket. A blown head gasket likely will cause air bubbles in the cooling system when the engine is running, and these will be visible in the header tank. A rare cause for the loss of coolant is a failure of the pressure cap on the header tank, which allows the coolant to slowly boil away.

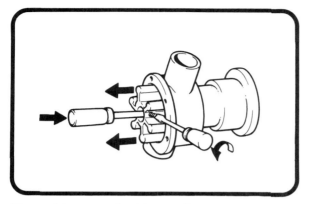

Figure 5-2. *Some flexible impellers also are secured to the shaft by a set screw. You must pry these out about 1/2 inch to undo the set screw. (Courtesy Volvo Penta)*

Overheating During Normal Operation.

Check the oil level. If a low oil level is causing the engine to overheat, expensive damage may be in the making. A partial seizure of one or more pistons may already be generating a tremendous amount of heat on the cylinder walls. More likely than a lack of oil, however, is a reduction in the raw-water flow through the cooling system.

The raw-water inlet screen (if fitted) on the outside of the hull may be blocked with a piece of plastic. Throttle down, put the boat in reverse, and throttle up. With any luck, the reverse propeller thrust will wash it clear. To confirm this, loosen a hose below the waterline and see if water flows into the boat. If the flow is restricted, check also for barnacles on the inlet

---

Troubleshooting Chart 5-2.
Overheating in Operation.

Check the raw-water overboard discharge. Is the flow less than normal? **NO**	**YES** Check for obstructions in the raw-water circuit (see troubleshooting chart 5-1). In addition, check the raw-water circuit for silting, scale and other partial obstructions (see the text).
Check the oil level. Is it low? **NO**	**YES** Refill with the correct grade and viscosity of oil.
Is the boat overloaded? (Check for a rope around the propeller; a heavily fouled bottom; adverse conditions; excessive auxiliary equipment; an oversized propeller.) **NO**	**YES** Reduce the loading.
Check the coolant level in the fresh-water circuit (if fitted). Caution: do not remove the header-tank pressure-cap when hot. Is the level low? **NO**	**YES** Refill and find the leak.
Is the fresh-water circuit air-locked? (With the header-tank pressure-cap off, check for signs of flow when the engine is running—you may have to wait for the engine to cool down and the thermostat to close.) **NO**	**YES** Break the hoses loose at the fresh-water pump and water heater (if fitted) and bleed off any air.
Is the thermostat operating incorrectly? (Remove and test as outlined in the text.) **NO**	**YES** Replace.
There is probably a mechanical problem (e.g. faulty fuel injection; a partial seizure)—see the text.	

screen, or in the water intake.

Check the raw-water filter. If you find a lot of silt, the heat exchanger (or engine itself on raw-water cooled boats) may be silted up. Feel the freshwater inlet and outlet pipes to the heat exchanger—if it is doing its job, there should be a noticeable fall in the temperature of the water leaving the heat exchanger. Many heat exchangers have removable end caps and can be rodded out with a suitable wooden dowel. If a refrigeration condenser is fitted in-line with the engine cooling circuit, any obstruction in the condenser will reduce the raw water flow to the engine.

Check for collapsed cooling hoses or poor pump performance. This is another reason that a high-set overboard discharge is useful, as the pump flow can be gauged at a glance. In certain instances, if the raw-water intake is not set low enough in the hull, a well-heeled sailboat can suck in air and air-lock the raw-water pump.

Where the raw water is injected into the exhaust (on both raw-water and heat-exchanger cooled engines) a relatively small nozzle is sometimes used to direct the water down the exhaust pipe and away from the exhaust manifold. If scale forms in the raw-water circuit, this nozzle is likely to plug, restricting the water flow.

If a boat has been operating in salt water then moves into fresh water, scale formed in salt water will swell and the engine will gradually overheat. On heat-exchanger engines with galvanized exhaust piping, check the pipes for partial blockages. On raw-water engines, first look at the water side of the exhaust manifold—a major descaling may be in order.

If the raw-water flow is normal, check any header tank as above, but *only after allowing the engine to cool.*

Perhaps the engine is overloaded (a rope around the propeller, an oversized propeller, a badly fouled bottom, too much auxiliary equipment). Maybe the ambient water temperature is higher than normal. A boat moving into the tropics may experience a 40 °F (22 °C) rise in water temperature, and engine temperatures may rise a little.

The thermostat may be malfunctioning (some raw-water engines don't have thermostats). It will be found under a bell-housing near the top and front of the engine. Take it out and try operating without it. This will make most engines run cool, but will cause

Figure 5-3. *Checking the operation of the thermostat. (Courtesy Volvo Penta)*

a few engines—e.g., some Caterpillars—to run hot and in this case should not be done. To test the thermostat, put it into a pan of water and heat it (see Figure 5-3). It should open between 165 °F and 185 °F (74 °C to 82 °C), except on some Caterpillars, which open as high as 192 °F (89 °C), and some raw-water engines, which open between 140 °F and 160 °F (60 °C and 71 °C).

Raw-water-cooled engines are likely to develop scale around the cylinders over time, especially if run above 160 °F (71 °C). The best defense against scale formation is to have adequate sacrificial zincs in the raw-water side of the cooling system, and *to change them as soon as they are partially eaten away.* This is one of the most important maintenance items on marine diesels. Descaling, when necessary, requires professional advice (see also page 93).

Faulty injection and injection dribble can cause late burning of the fuel, which will heat up the cylinders at the end of the power stroke and would normally be accompanied by black smoke. This heat is not converted into work and must be removed by the cooling system, leading to higher temperatures. Scale, or partial seizures (see below), can generate hot spots that cause pockets of steam to build up in the cylinder or head. These can sometimes air-lock cooling passages, the cooling pump, the heat exchanger, or the expansion tank, especially if the piping runs have high spots where steam or air can gather. (Note that where a domestic water heater is plumbed into the engine cooling circuit, as most are, there is frequently a high spot in the tubing that can cause an air-lock.) As noted above, a blown head gasket is likely to cause bubbles in the cooling system.

Finally, problems with temperature gauges are rare. If suspected, consult the section on "Engine Instrumentation" (page 85).

Smoke

The exhaust of a diesel should normally be perfectly clear. The presence of smoke often can point to a problem in the making, and the color of the smoke can be an even more useful guide.

Black Smoke. Black smoke is the result of unburned particles of carbon from the fuel blowing out of the exhaust. On many older engines, any attempt to accelerate suddenly will generate a cloud of black smoke as the fuel rack opens and the engine slowly responds. Once the engine reaches the new speed setting, the governor eases off the fuel rack and the smoke immediately ceases. This smoke is indicative of a general engine deterioration—the compression is most likely falling, the injectors need cleaning, and the air filter should be changed. If the engine is otherwise performing well, you have no immediate cause for concern, but the engine is serving notice that a thorough service is overdue.

If smoking persists when the load is eased off, the engine is crying out for immediate attention. The following are likely causes for the smoke:

- Obstruction in the airflow through the engine. This results in insufficient air entering the cylinders for proper combustion to take place. Likely causes are a dirty air filter, or restrictions in the air inlet ducting, or a high exhaust back pressure (see page 79). Turbocharged engines in particular are sensitive to high back pressure. The turbocharger slows down, and as a result pushes less air into the engine than it requires, leaving fuel unburned. A dirty or

Troubleshooting Chart 5-3.	
Smoke	
Black	**Blue**

Obstruction in the airflow:
    Dirty air filter
    Defective turbo/supercharger
    High exhaust back pressure
	Worn or stuck piston rings

	Worn valve guides

Excessively high ambient air
    temperature
	Turbo/supercharger problems:
	    worn oil seals
	    plugged oil drain

Overload:
    Rope around the propeller
    Oversized propeller
    Heavily fouled bottom
    Excessive auxiliary equipment
	Overfilled oil-bath type air
	    filter

	High crankcase oil level/ pressure
Defective fuel injection	
White	
Lack of compression Water in the fuel Air in the fuel Defective injector Cracked cylinder head/leaking head gasket	

defective turbocharger will have the same effect (see page 84). Many engines on auxiliary sailboats are tucked away in little boxes. As often as not these are fairly well sealed to cut down on the noise levels. Unless they are adequately vented, *this can strangle the engine*, particularly in hot climates, where the air is less dense, and at higher engine loadings.

- Overloading of the engine. The governor reacts by opening the fuel control lever until more fuel is being injected than can be burned with the oxygen that's available. This improperly burned fuel is emitted as black smoke.

It is an unfortunately common practice to fit the most powerful propeller that the engine can handle in optimum conditions—a lightly loaded boat with a clean, drag-free bottom in smooth water. This practice exaggerates the performance of the boat under power, but overloading results under normal operating conditions. Adding various auxiliary loads, such as a refrigeration compressor and a high output alternator, often compounds the problem because they aren't taken into account in the overall power equation. Black smoke on new boats should always lead you to suspect an overloaded engine caused by the wrong propeller or too much auxiliary equipment. (Matching engines to their loads is dealt with in more detail in Chapter 9.) Overloading can also arise from wrapping a rope around the propeller.

Apart from the likelihood of a smoky exhaust, overloading is liable to cause localized overheating in the cylinders, which will definitely shorten engine life and could lead to an engine seizure.

- Defective fuel injection. Dirty, plugged, or worn injector nozzles can cause inadequate atomization of the fuel, improper distribution of the fuel around the cylinders, or result in a dribble after the main injection pulse. All can lead to unburned fuel and black smoke. If an engine is not overloaded, and the air flow is unobstructed, poor injection is the number one suspect for black smoke.
- Excessively high ambient air temperatures (e.g., in a hot engine room on a boat operating in the tropics). The density, thus the weight, of the air entering the engine will be reduced, leading to an insufficient air supply, especially at high engine loadings.

Blue Smoke. Blue smoke arises from the burning of the engine's lubricating oil. This oil can only find its way into the combustion chambers by making it up past piston rings, down valve guides and stems, or in through the air inlet from leaking supercharger or turbocharger seals, from an overfilled oil-bath air filter, or perhaps from a crankcase breather. A plugged oil drain in the turbocharger will also cause oil to leak into the compressor housing and enter the air inlet.

Engines that are repeatedly idled or run at low loads do not become hot enough to fully expand the pistons and piston rings. These then fail to seat properly, and oil from the crankcase finds its way into the combustion chamber. In time, the cylinders become *glazed* (very smooth), while the piston rings get gummed into their grooves, allowing more oil past. Oil consumption rises and compression declines. Blow-by down the sides of the pistons raises the pressure in the crankcase and blows an oil mist out of the crankcase breather. Carbon builds up on the valves and valve stems and plugs the exhaust system. Valves may jam in their guides and hit pistons. Idling and low-load running will substantially increase maintenance costs, including major overhauls, and shorten engine life.

White Smoke. White smoke is caused by water vapor in the exhaust or by totally unburned, but atomized, fuel. The former is symptomatic of dirty fuel, or possibly a leaking head gasket, cracked head, or cracked cylinder allowing water into the combustion chamber. The latter generally indicates that one or more cylinders are failing to fire. Air in the fuel supply will, on occasion, cause misfiring with little puffs of white smoke. Unfortunately, it is generally not possible to distinguish any of these conditions on a boat because of the water normally present in the exhaust which is injected from the cooling system.

Knocks

Diesel engines make a variety of interesting noises. Each of the principal components creates its own sound, and a good mechanic can often isolate a problem simply by detecting a specific *knock* coming out of the engine.

In addition to the injector creak already mentioned ("Bleeding a fuel system," page 61), at any point on the valve cover you can hear the light tap, tap, tap of the rocker arms against the valve stems. (The adjusting screws that are used to set valve-lash clearance—see Chapter 7—are known as *tappets*.) With practice, anyone can pick up the note of individual tappets and get a pretty good idea if valve clearances are correct. Water pumps, camshafts, and fuel pumps have their characteristic note, as do most other major engine components.

The symphony, however, is frequently garbled by a variety of fuel and ignition knocks. Differences in the rate of combustion can cause noises that are almost indistinguishable from mechanical knocks, especially on 2-cycle diesels. But if the engine is run at full speed and then the governor control lever (the throttle) shut down, a fuel knock will cease at once, whereas a mechanical knock will probably still be audible, albeit not as loudly as before because the engine is now merely coasting to a stop. Knocks which gradually get louder over the life of an engine (especially if accompanied by slowly declining oil pressure when the engine is hot) are almost certainly mechanical knocks.

Fuel-side Knocks.
Some fuel knocks are quite normal, especially on initial start-up. Remember that diesels are much noisier than gasoline engines and have a characteristic clatter at idle, especially when they are cold. The owner of a diesel will have to become accustomed to these noises in order to detect and differentiate out-of-the-ordinary fuel knocks. These can have several causes:

- Poor-quality fuel (low cetane rating; dirt or water in the fuel). The fuel is slow to ignite and builds up in the cylinder. But then the heat generated by the early part of combustion causes the remaining (and now excessive) fuel to burn all at once. The sudden expansion of the gases causes a shock wave to travel through the cylinder at the speed of sound. You will hear and feel this as a distinct knock (see Figure 5-4). It is known as *detonation*.

- Faulty injector nozzles. These can produce a result similar to that just described. The fuel is not properly atomized and as a result initial combustion is delayed. The fuel builds up in the cylinder then a sudden flare-up occurs.

- Injection timing too early. This causes the fuel to start burning while the piston is still traveling up on its compression stroke. The piston is severely stressed as the initial combustion attempts to force the piston back down its cylinder before the crankshaft has come over top dead center (TDC). Timing problems should not be encountered in normal circumstances.

- Oil in the inlet manifold. On supercharged and turbocharged engines, leaking oil seals will sometimes allow oil into the inlet manifold. The oil is then sucked into the cylinders and can cause detonation. In extreme cases, enough oil can be drawn in to cause engine *runaway*—the engine speeds up out of control and will not shut down when the fuel rack is closed (see below).

Mechanical Knocks.
The more common mechanical knocks arise from:

- Worn piston pin or connecting rod bearings. As the piston reaches top dead center (TDC) or bottom dead center (BDC), its momentum carries it one way while the crankshaft is moving the other. Any play in the bearings will result in a distinct noise that varies with engine speed. A worn connecting rod bearing knocks more loudly under load. By testing a variety of points along the crankcase and block, you can sometimes isolate the specific connecting rod at fault. A good trick is to touch the tip of a long screwdriver to the engine while placing your ear to the handle—it will act like a stethoscope, amplifying internal sounds. Be sure to tie up long hair and keep clear of moving belts, flywheels, etc. . . . .

- Worn pistons *slap* or *rattle* in their cylinders. This is more audible at low loads and speeds, particularly when idling.

- Worn main bearings *rumble* rather than knock. Engine vibration increases, especially at higher engine speeds.

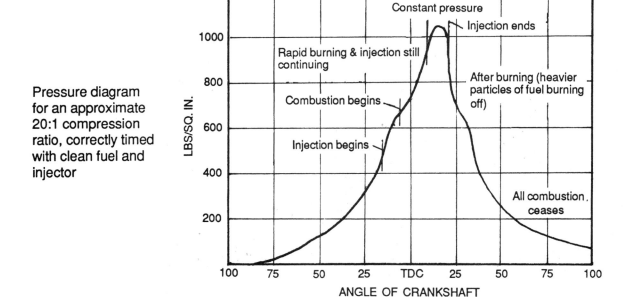

Pressure diagram for an approximate 20:1 compression ratio, correctly timed with clean fuel and injector

Figure 5-4. *Fuel combustion pressure curves.*

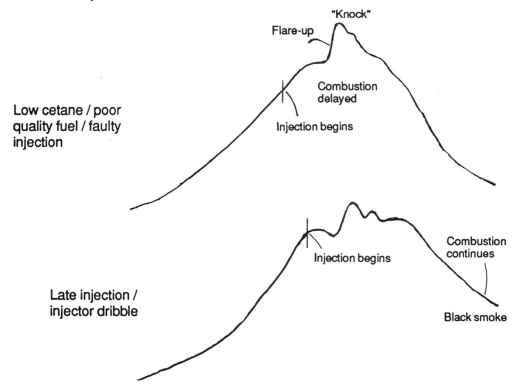

Low cetane / poor quality fuel / faulty injection

Late injection / injector dribble

Misfiring Cylinders

Diesels idle unevenly, especially those with jerk-type injection pumps, due to the difficulty of accurately metering the minute quantities of fuel required at slow speeds. This is not to be mistaken for misfiring, which will be felt and heard as rough running at all speeds.

Misfiring may be rhythmical or erratic. The former indicates that the same cylinder(s) are misfiring all the time; the latter means that cylinders misfire randomly. Rhythmical misfiring is caused by a specific problem with one or more cylinders, such as low compression or faulty fuel injection. If it occurs on start-up then clears up once the engine warms, it is almost certainly due to low compression. The air in the cylinder is not initially reaching ignition temperature, but as the engine warms the air gets hotter until the cylinder fires.

The guilty cylinder(s) of a rhythmical misfire can be tracked down by loosening the injector nut on each injector in turn (with the engine running) until fuel spurts out. If the engine changes its note or slows down, the cylinder was firing as it should, and you can retighten the nut. If no change occurs, the cylinder is misfiring, and you need to investigate further. (Of course, if no fuel spurts out, we know why it's not firing!)

If the engine has internal fuel lines, loosen the delivery pipes at the external fittings on the cylinder head so that fuel doesn't run down into the crankcase. Detroit Diesels are a little different. First of all they have no suitable external fuel lines, and even if they did, loosening a fuel line would cause large amounts of fuel to flow out. You can, however, remove the valve cover and hold down the actuating rocker on individual injectors to disable them (see Figure 5-5). If this slows the engine and changes its note, you know that this cylinder is firing OK. If there is no change, the cylinder is missing. If you find a problem with a Detriot Diesel that you are unable to solve, in a dire emergency you can take the offending injector out of service by disconnecting the fuel lines from the injector, turning the injector around in the cylinder

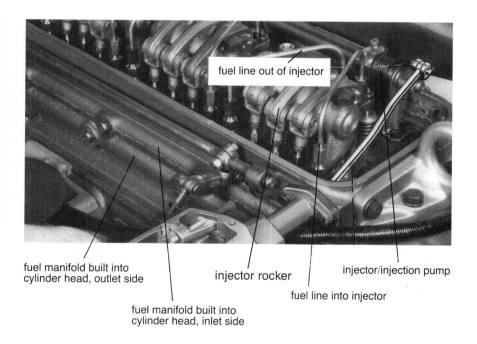

fuel line out of injector

fuel manifold built into cylinder head, outlet side

injector rocker

injector/injection pump

fuel line into injector

fuel manifold built into cylinder head, inlet side

Figure 5-5. *A common rail fuel system. (Courtesy Detroit Diesel Corp.)*

head, and reconnecting the fuel lines. You can then continue to run the engine on the remaining cylinders until you find help.

Erratic misfiring on all cylinders is the result of a general engine problem, frequently contaminated fuel. If the misfiring is more pronounced at higher speeds and loads, in all probability the fuel filter is plugged. If it is accompanied by black smoke, the air filter is probably plugged, or there is some other problem with the airflow (carbon in the exhaust, defective turbocharger, etc.). In rare cases, poorly seated lift pump valves or a damaged diaphragm may be leading to fuel starvation.

Seizure

Seizure of the pistons in their cylinders is an ever-present possibility any time serious overheating occurs, or the lubrication breaks down. Overheated pistons expand excessively and jam in their cylinders. An engine experiencing a seizure *bogs down*—that is to say, fails to carry the load, slows down progressively, probably emits black smoke, and becomes extremely hot. If you don't take steps immediately to deal with the situation, total seizure—the engine grinding to a halt and locking up solidly—is not far off.

Partial seizures of individual pistons can occur as a result of uneven loading, causing overheating in the overloaded cylinder. Sometimes a serious case of injector malfunction and dribble, particularly on a direct-combustion engine, results in a stream of diesel washing the lubrication off a part of the cylinder wall, which leads to a partial seizure of the piston. I have even seen a new engine seized absolutely solid, while it was shut down, by the differential contraction of the pistons and cylinders in a spell of extremely cold weather.

If you detect the beginning of a partial engine seizure, the correct response is not necessarily to shut the engine down immediately—as it cools the cylinders are likely to lock up solidly on the pistons. The load should be instantly thrown off and the engine idled down as far as possible for a minute or so to give it a chance to cool off. This action assumes that the seizure is not due to the loss of the lubricating oil or cooling water. In either of these situations, you have no choice but to shut down as soon as possible.

Poor Pick-up

Poor pick-up, or a failure to come to speed, is most likely to be a result of one or more of the following:

- Insufficient fuel caused by a plugged filter or nearly empty tank;
- Clogged air filter, in which case the engine is likely to be emitting black smoke; (*You should have noticed that this is about the fifth time that plugged filters have been mentioned in connection with various problems! This is no accident. Routine attention to filters is neglected time and time again.*)
- Supercharger or turbocharger malfunction—black smoke is likely;
- Overloading due to an improperly matched propeller, a heavily fouled bottom, too much auxiliary equipment, or perhaps a rope around the propeller;
- Excessive back-pressure in the exhaust (see below);
- Low compression, perhaps resulting in misfiring;
- Dirty or clogged injector nozzles;
- Injection pump plungers leaking due to excessive wear;
- Too much friction—a partial seizure is under way.

Sticking or Bent Valve Stems

Many marine diesels, particularly those fitted to auxiliary sailboats, are run at idle and low operating temperatures for prolonged periods. As noted above ("Blue smoke") this leads to carbon formation throughout the engine. Using non-detergent oils in the engine can have the same effect, or neglecting oil and filter changes. Valve stems and guides become coated with carbon, and after a while, valves are likely to jam in their guides in the open position. When the piston comes up on its exhaust or compression stroke, it may hit the open valve, knocking it shut or bending the valve stem and damaging the piston crown.

As a temporary measure, sticking valves often can be relieved by lubricating the stem with kerosene and turning the valve in its guide to loosen it. At the earliest opportunity, the cylinder head will have to be removed, and the valves, guides, etc., thoroughly decarbonized (see Chapter 7).

Engine Runaway

Engine runaway occurs as a result of oil, or other combustibles (e.g., propane from a leaking gas cylinder that has not been installed in a properly vented compartment) being sucked into the engine through the air-inlet manifold. Oil can come from an overfilled oil-bath air cleaner, an overfilled crankcase spewing oil out of its breather, failed turbocharger or supercharger oil seals, or arise as a result of a plugged oil drain on a turbocharger forcing oil into the compressor housing.

Runaway is more prevalent on 2-cycle diesels than 4-cycles, and that's why Detroit Diesels have the emergency air flap that cuts off all air to the engine and strangles it. In the absence of an emergency air flap, the only way to stop runaway is to cut off the oxygen supply to the engine. The most effective way to do this is to aim a CO_2 (carbon dioxide) fire extin-guisher into the air inlet or to jam a blanket or boat cushion up the inlet, *making sure you keep your hands out of the way.* If not stopped, runaway can continue until the oil supply is used up. If the oil is coming through leaking supercharger/turbocharger oil seals, it will normally be coming from the engine's oil supply, and eventually the engine will seize.

High Exhaust Back Pressure

Exhaust back pressure can be checked with a very sensitive pressure gauge designed to measure in *inches of mercury*, or *inches of water*, rather than pounds per square inch (psi). (See Table 5-1 for conversion of one to the other.) It can also be checked with a homemade *manometer*. This can be nothing more than a piece of clear plastic tubing of any diameter greater than $1/4''$, fixed in a U-shaped loop to a board about 4 feet long. The board is marked off in inches (see Figure 5-6).

Set up the board on end and half fill the tubing with water. Connect one side of the tubing to a fitting

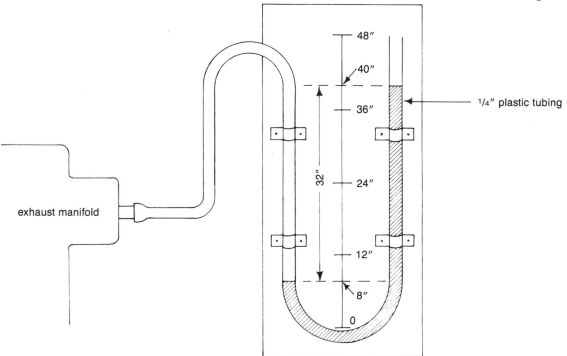

Figure 5-6. *A simple homemade manometer to check exhaust back pressure.*

on the exhaust as close to the exhaust manifold as possible (but 6" to 12" *after* a turbocharger). Leave open the other end of the tubing. If the manifold has no suitable outlet to make the connection, drill an $^{11}/_{32}$" hole, tap this for a $^1/_8$" pipe fitting (standard pipe thread), and screw in an appropriate fitting. When finished, remove the fitting and fit a $^1/_8$" pipe plug.

Note the level of the water in the tube with the engine at rest. Then crank the engine and *fully load* it. *This is important—if necessary, tie the boat off securely to a dock, put it in gear, and open the throttle.* The exhaust back pressure will push the water down one side of the tubing and up the other. The difference between the two levels, measured in inches, is the back pressure in *inches of water column*. On naturally aspirated engines and Detroit Diesels, it should not exceed 40" of water (3" Hg); on turbocharged engines, including turbocharged Detroit Diesels, 20" of water (1.5" Hg). However, it is frequently hard to keep an exhaust with a water−lift-type muffler within these tolerances (see page 192).

Table 5-1

To convert	Into	Multiply by
Psi	Hg"	2.036
Hg"	Psi	0.4912
Psi	H_2O"	27.6776
H_2O"	Psi	0.03613
H_2O"	Hg"	0.07355
Hg"	H_2O"	13.5962

The most likely causes of high exhaust back pressure are:
- A closed, or partially closed, sea valve on the exhaust exit pipe;
- Too small an exhaust pipe, too many bends and elbows, too great a lift from a water-lift muffler to the exhaust exit (see page 192), or a kink in an exhaust hose;
- Excessive carbon formation in the exhaust system;
- In wintertime, frozen water in a water-lift-type muffler at initial start-up.

If exhaust back pressure is excessive, check the exhaust for carbon formation or obstructions. If the exhaust is clear, it will need modifying to reduce the back pressure (see Chapter 9).

The exhaust can reveal in other ways a surprising amount of information about the operation of an engine. One of the very best methods for monitoring the performance of an engine, used on all large diesels, is an exhaust pyrometer fitted to each cylinder. These measure the temperature of the exhaust gases as they emerge from the cylinders. Variations in temperature from one cylinder to another show unequal work due to faulty injection, blow-by, etc., and should never exceed + or − 20°F (11°C).

On occasion, exhaust pyrometers are offered as an option on smaller diesels. For those inclined to do all their own troubleshooting and maintenance, they are a worthwhile investment. This is especially so with today's higher-revving, hotter-running, and more highly-stressed engines. High exhaust temperatures on any cylinder will sharply decrease engine life. The additional cost of a pyrometer installation can easily be paid for in a better-balanced and longer-lived engine.

High Crankcase Pressure

Smoke blowing out of the crankcase breather or dipstick hole is an indication of high crankcase pressure. This condition is likely to develop slowly, generally as a result of poorly seating or broken piston rings, which allow the gases of combustion to blow-by the pistons into the crankcase. Other causes may be a crack or hole in a piston crown (top), a blown head gasket, or excessive back pressure. A compression test on the cylinders will reveal if just one cylinder is blowing-by, or all.

On turbocharged engines, air pressure will blow down the oil drain lines into the crankcase if the oil seals begin to give out, whereas on Detroit Diesels a damaged gasket between the supercharger and the block has the same effect.

Water in the Crankcase

A certain amount of water can find its way into the crankcase from condensation of the steam formed during combustion, but appreciable quantities can only come from the cooling system. The sources are strictly limited: water siphoning in through exhaust

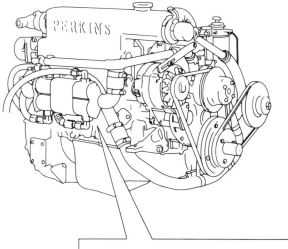

Figure 5-7. *Oil coolers increase engine life. Cooling water passes through small-diameter copper tubing inside the oil reservoir, drawing heat from the oil. This unit from Perkins cools engine and transmission oil.*

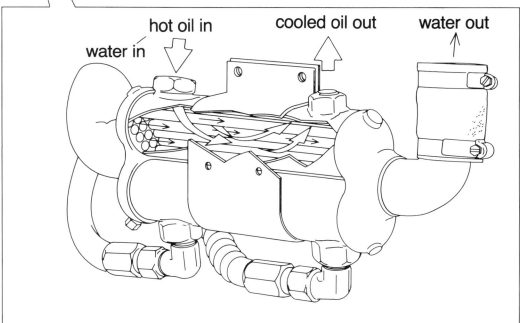

valves from a faulty water-cooled exhaust installation (see Chapter 9), leaks around injector sleeves (where fitted), a leaking cylinder head gasket, a cracked cylinder head or liner (or one with a pinhole caused by corrosion from the water-jacket side), a leaking O-ring seal at the base of a wet liner, or corrosion in an oil cooler.

The cooling tubes on oil coolers with raw water circulating on the water side (as opposed to water from an enclosed engine cooling circuit) are especially prone to damage. The combination of heat, salt water, and dissimilar metals is a potent one for galvanic corrosion. All too many oil coolers are made of materials unsuited to the marine environment (e.g., brass). Oil coolers are expensive and often hard to find. Before starting on a long cruise, make sure you have a proper marine-grade cooler (bronze and cupronickel—see Figure 5-7—*which is adequately protected with sacrificial zinc pencil anodes. These zincs should be checked every month or so and replaced when only partially eaten away* (no more than 50%). Some engines in the tropics will consume their zincs in as little as 30 days—be sure to bring a good supply of spares (for more on zincs, see page 97).

When an oil cooler tube fails during engine operation, oil is likely to be pumped into the cooling system. On a raw-water oil cooler, the oil will show itself as an overboard slick. On a freshwater oil cooler, the oil will appear in the header tank. When the engine is at rest, water from the cooling system may find its way into the oil side of the cooler and siphon into the crankcase.

If fittings can be found to by-pass the oil and water sides of a failed oil cooler, the engine can still be run until you find a replacement cooler, but only at low loadings and only after changing the engine oil and filter (see below). Keep a close eye on the oil-pressure and engine-temperature gauges.

You may encounter one condition that is sometimes mistaken for a water leak into the engine but in fact is not. This is condensation in the valve cover, leading to emulsification of the oil in the valve cover (it goes gooey and turns a creamy color), and rusting of the valve springs and other parts in the valve train. This happens from time to time when the owner periodically runs the engine for relatively short periods of time to charge a battery, or just to make sure it is still working. The engine never warms up properly, but it generates enough heat to create condensation in the valve cover.

If an engine is started, it should be allowed to run long enough to thoroughly warm it up and allow the hot engine oil to cook out all the moisture in it. What is more, the engine should be given some work to do. As noted previously, lightly-loaded diesels run erratically (due to the difficulty of accurately metering the minute quantities of fuel required at each injection), and tend to carbon up. If possible, tie off the boat firmly, put the engine in gear, and open the throttle a little.

Flushing the Lubricating System. If an oil cooler is cooled from an enclosed engine-cooling circuit and if the engine has ethylene glycol in the cooling circuit (as it should have), a failed oil cooler tube will allow the glycol into the crankcase (as may a blown head gasket). When mixed with oil, *ethylene glycol forms a varnish that coats bearing surfaces and may lead to engine seizure*. It also forms a soft, sticky carbon sludge that will adhere to oil passages and coat

valves, causing them to stick.

Drain the oil, change the filter, and flush the engine with a solution of two parts Butyl Cellosolve to one part SAE 10 engine oil at the earliest possible opportunity. Fill the engine to its normal level, start it, and run it at a fast idle (with no load) for 30 minutes. Thoroughly drain it, change the filter, refill it with SAE 10 oil, and run it for another 10 minutes. Drain it and change the filter once again, and refill with its normal oil. Even with this procedure, it is very hard to remove the varnish.

Flushing the Cooling System. If oil has entered an enclosed cooling system, it will show up in the header tank. Drain the cooling system and remove the thermostat (on most engines, but not where this produces a high temperature—e.g., some Caterpillars). Flush the system with fresh water to remove as much of the oil as possible. Refill it and run the engine at a fast idle until it is warm. Remove the pressure cap from the header tank (take care!) and add a cup of non-foaming detergent (e.g., automatic dishwasher soap—use approximately 2 ounces of soap per gallon of water). Run the engine another 20 minutes. If there is still undissolved oil on the surface of the water, add another cup of soap and run another 10 minutes. Drain, and refill. If oil is still present, repeat the procedure. When the system is clean, replace the thermostat (if removed), and refill with water and antifreeze.

Low Oil Pressure

Low oil pressure is a serious problem, but occurs infrequently. Many people, confronted with low oil pressure, assume that the gauge or warning light is malfunctioning and ignore the warning. Given the massive amount of damage that can be caused by running an engine with inadequate oil pressure, this is the height of foolishness. *Any time low oil pressure is indicated, immediately shut down the engine, find the cause, and fix it.*

The problem is likely to be one of the following:
- Lack of oil—the most common cause of low oil pressure and the least forgivable;
- The wrong grade of oil in the engine—may lead to the viscosity being too low;
- Overheating, lowering the viscosity of the oil

even though the correct grade is in the engine;

- Diesel dilution of the oil—this can occur on engines with internal fuel lines. The oil level will rise; once enough diesel has found its way into the oil to lower the pressure to a noticeable extent, it will be possible to smell the fuel in the oil if you take a sample from the dipstick and rub it on your fingers.

- Worn bearings. The oil pump feeds oil under pressure through holes drilled in the crankcase (called *galleries*) and through various pipes to all the engine bearings. Oil squeezes out between the two surfaces of the bearings. The pump has a large enough capacity to circulate more oil than is needed to maintain the correct pressure in the system, and the surplus bleeds back to the sump via a *pressure relief valve*. In time, wear on the bearings allows the oil to flow out more freely. Eventually the rate of flow is greater than the pump can sustain at the normal operating pressure, and the pressure declines.

  Worn bearings do not, as a rule, develop overnight. A very gradual decline in oil pressure occurs, especially at low engine speeds when the engine is hot. There will also be a gradual increase in engine noise and knocks. By the time a significant loss of oil pressure occurs, many other problems are likely to be evident, and a major rebuild is called for. *Any rapid loss of oil pressure accompanied by a new engine knock indicates a specific bearing failure that needs immediate attention.*

- Oil pressure relief valves sometimes malfunction, generally through a broken spring or because of trash on the valve seat (especially if proper oil change procedures have been ignored). In this case, the oil is vented directly back to the sump, with a consequent loss in pressure.

  Problems with pressure relief valves are rare but are simple to check. Almost invariably, the pressure relief valve is screwed into the side of the block somewhere or fitted to the oil filter mount and can be easily removed, disassembled, cleaned, and put back. The spring is liable to be under some tension, so take care when taking the valve apart. After cleaning,

reset the spring's tension to maintain the manufacturer's specified oil pressure. Run the engine until it is warm and check the oil pressure. If it is low, shut down the engine and tighten the relief valve spring a little or add more shims. If no amount of tensioning on the spring brings the oil pressure up to the manufacturer's specifications, the problem lies elsewhere.

- Oil pumps rarely, if ever, give out, as long as the oil is kept topped up and clean and the filter is changed regularly. Over a long period of time, wear in an oil pump may produce a decline in pressure, but not before wear in the rest of the engine creates the need for a major rebuild. At this time, the oil pump should always be checked.

- A clogged oil cooler will reduce oil pressure. Clogging can arise from long hours of low load running and idling (which generates a carbon sludge within the engine), coupled with a failure to change the oil at the specified intervals (or using non-detergent oil in the engine). In this case, other oil galleries are likely to be plugged with sludge and serious engine damage is probable. The oil side of most oil coolers cannot be flushed—the oil cooler will need replacing.

- A well-heeled sailboat will sometimes cause the oil pump suction line to come clear of the oil in the pan (sump), allowing the pump to suck in a slug of air. The oil pressure will momentarily drop, generally with a sudden, alarming clatter from the engine. This is especially likely to happen if the oil level is a little low. Check the level, top up as necessary, and put the boat on a more even keel.

- The failure of an external oil line or gasket (e.g., to an oil cooler) will cause a sudden, potentially catastrophic loss of oil and pressure. The engine is likely to suddenly clatter loudly. It must be shut down immediately. There will be oil all over the engine room! Less easy to spot is the loss of oil that accompanies a corroded cooling tube in an oil cooler. The oil will be pumped out of the exhaust (or into the header tank), and sometimes water will enter the crankcase (see above).

- The oil pressure gauge is unlikely to malfunction. The most reliable ones in the marine environment are the older mechanical gauges that have an oil line connected from the engine block directly to the gauge. However, if the oil-sensing line fails, the engine will pump its oil out of the broken line, spraying it all over the engine room. Most modern engines have electronic pressure sensing devices on the engine block (the sender), with a wire to an electronic gauge. If salt water finds its way into the electrical circuit, the gauge is likely to malfunction (see below for troubleshooting engine instruments).

Rising Oil Level

If the oil in the crankcase starts to rise, it is likely to be because water has entered from the cooling system, or diesel from a ruptured or leaking internal fuel line (if the engine has internal fuel lines). Another source is from blown seals between the engine and some hydraulic transmissions, since the oil pressure in the transmission is generally higher than in the engine.

High Oil Consumption

Is the engine making a lot of blue smoke? If so, *it is burning the oil*, most likely because of gummed up or broken piston rings. The oil also may be entering the combustion chambers down worn valve guides, through failed seals or a plugged oil drain in a turbocharger or supercharger, or from the engine breather. (Maybe the crankcase is overfilled with oil; otherwise there is likely to be excessive crankcase pressure as a result of blow-by down the sides of pistons.)

If there is no blue smoke, check for external leaks from oil lines and gaskets. If the engine has a raw-water oil cooler, check the overboard water discharge for an oil slick as a result of a holed tube in the oil cooler. If the oil cooler is fresh-water-cooled, look for oil in the header tank. (Remember, do not remove the cap when hot.)

If an engine appears *to stop burning oil*, it may simply be due to overfilling, or misreading the dipstick, but it may also be the result of a slow water leak, or a slow diesel leak (if there are internal fuel lines), into the oil, keeping up with oil consumption. It needs to be checked out immediately.

Inadequate Turbocharger Performance

Poor turbocharger performance will cause symptoms similar to those caused by a plugged air filter: reduced power, overheating, and black smoke. Be sure to check out the airflow through the engine (including looking for obstructions in the exhaust, and dirt plugging the fins on any intercooler or aftercooler) before turning your attention to the turbocharger.

Spinning at up to 120,000 r.p.m., a turbocharger's blade tips can exceed the speed of sound; temperatures run as high as 1200°F (650°C)—hot enough to melt glass. The degree of precision needed to make all this possible means that *turbochargers are strictly items for specialists*.

A turbocharged engine should never be raced on initial start-up—the oil needs time to be pumped up to the bearings. Similarly, never race the engine before shutting it down; the turbine and compressor wheel will continue to spin for some time but without any oil supply to the bearings.

Clean oil is critical to turbocharger life—the bearings will be one of the first things to suffer from poor oil change procedures. Many engines have a bypass valve fitted to the oil filter so that if the filter becomes plugged, *unfiltered* oil will circulate through the engine. If the filter is neglected for long, the turbocharger will soon be damaged. Note that some turbochargers also have their own oil filter, which must be changed at the same time as the engine oil filter.

Any loss of engine oil pressure, such as from a low oil level or the wrong grade of oil being used in the engine, will also threaten the turbocharger. When a turbocharger is under load, insufficient oil for as little as five seconds can cause damage. Damage to bearings will allow motion in the shaft, permitting the turbines to rub against their housings.

A dirty or damaged air filter, or leaks in the air-inlet ducting, will allow dirt particles into the turbocharger, which will erode the compressor wheel and turbine. The resulting loss of performance, and imbalance, will lead to other problems.

Before condemning a turbocharger you can make the following tests:

1. Start the engine and listen to it. If a turbocharger is cycling up and down in pitch, there is probably a restriction in the air inlet (most likely a plugged filter). A whistling sound is quite likely produced by a leak in the inlet or exhaust piping.

2. Stop the engine, *let the turbocharger cool*, and remove the inlet and exhaust pipes from the turbine and compressor housings (these are the pipes going into the *center* of the housings). This will give a view of the turbine and compressor wheels. Use a flashlight to check for bent or chipped blades, erosion of the blades, rub marks on the wheels or housings, excessive dirt on the wheels, or oil in the housings. The latter may indicate oil seal failure, but first check for other possible sources, such as oil coming up a crankcase breather into the air inlet, oil from an overfilled oil–bath-type air cleaner, or a plugged oil drain in the turbocharger, causing oil to leak into the turbine or compressor housings.

3. Push in the wheels and turn them to feel for any rubbing or binding. Do this from both sides.

If these tests reveal no problems, the turbocharger is probably OK. If it failed on any count (except dirty turbine or compressor wheels—for cleaning, see page 102), it should be removed as a unit and sent in for repair.

Problems with Engine Instrumentation

Some engines still have thermometer-type temperature gauges and mechanical pressure gauges and tachometers. All will have some kind of a metal tube from the engine block to the back of the gauge. Gauge failure is normally self-evident—the gauge sticks in one position. Temperature gauges and their sensing bulbs have to be replaced as a complete unit; oil pressure gauges may just have a kinked sensing line; tachometers may have a broken inner cable.

Most engines today use electronic instruments comprising a sending unit on the engine block transmitting a signal to a gauge, warning light, or alarm (see Figure 5-8). Detecting problems is not quite so straightforward.

Ignition Warning Lights. Most alternators require an external DC power source to *excite* the alternator before it will start generating power. An ignition warning light is installed in the *excitation line* to the alternator. When the ignition is turned on, current runs from the battery down this line to the alternator, causing the light to glow. When the engine fires up, and the alternator begins to put out, the light is extinguished (see Figure 5-9). If the light fails to come on when the ignition is switched on, check the bulb first. If it is OK, the most likely problem is a break in the wire to the lightbulb, or in the excitation line running to the alternator, or an electrically poor connection. If the light comes on and stays on after the engine is running, the alternator is almost certainly not putting out (for troubleshooting alternators and alternator circuits, see *Boatowner's Mechanical and Electrical Manual*).

Other Warning Lights and Alarms. These use a simple switch. Positive current from the battery is fed via the ignition switch to the alarm or warning light and from there down to a switch on the engine block. If the engine reaches a preset temperature, or oil pressure drops below a preset level, the switch closes and completes the circuit.

Most switches (sending units) are the earth-return type—i.e., grounded through the engine block (see Figure 5-10). However, some are for use in insulated circuits, in which case they have a separate ground wire. An insulated return is preferred in marine use, especially when connected to gauges (as opposed to warning lights or alarms) since any time the ignition is on a small current is flowing through a gauge circuit, which may contribute to stray-current corrosion. (A warning light or alarm circuit differs in that the circuit conducts only when the light or alarm is activated, and therefore in normal circumstances will not contribute to stray-current corrosion.)

Many sending units incorporate both an alarm and a variable resistor, which connects to a gauge. In this case there will be two wires on an earth-return unit (see Figure 5-11), and three on an insulated unit.

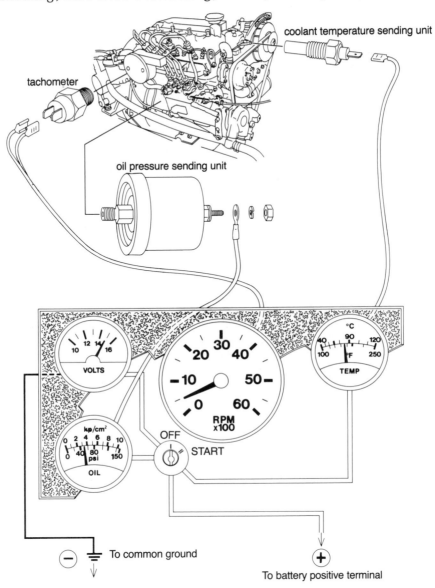

Figure 5-8. *How the typical instrument panel receives information from the engine.*

Figure 5-9. *Voltage regulator with excitation circuit. Closing the switch provides initial field current. The battery discharges through the ignition light into the field winding, lighting the lamp. When the alternator's output builds, the cutout points close. Now, the system has equal voltage on both sides of the ignition warning light, and the light goes out.*

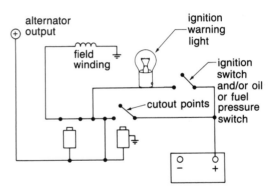

Warning light or alarm

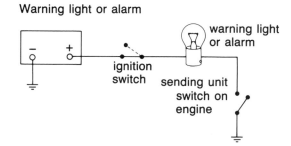

Figure 5-10. *A simple warning light or alarm circuit.*

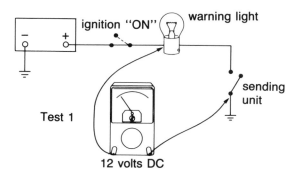

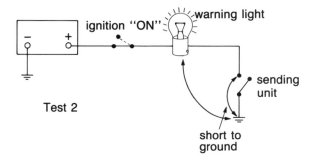

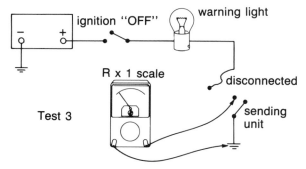

Figure 5-12. *Testing a warning light or alarm circuit.*

W = warning light or alarm wire
G = gauge wire

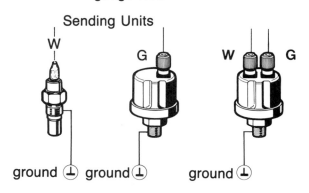

Figure 5-11. *Sending units for alarms and gauges. Left: Simple sensor with a warning contact (W). This type is used, for example, in an oil-pressure warning light system. Middle: Sensor with measuring contact (G) for a display instrument. This provides continuous value for a given operational condition. Right: Sensor with W and G contacts for a continuous instrument display and for warning when a critical value has been reached. (Courtesy VDO)*

If problems are suspected with an *alarm* or *warning light* (for *gauges*, see below), turn on the ignition switch and:

1. Test for 12 volts between the alarm or light positive terminal and its negative terminal (or a good ground on the engine block if there is no negative connection—see Figure 5-12). No volts: the ignition circuit is faulty. 12 volts: proceed to the next step.

2. To test the alarm or light itself, disconnect the wire from the sending unit and short it to a good ground. The alarm or light should come on. If not, make the same test from the second terminal on the back of the alarm or light (the one with the wire going to the sending unit) to a good ground. No response: the alarm or light is faulty. Response: the wire to the sending unit is faulty.

3. If the alarm or light and its wiring are in order, the sending unit may itself be shorted (the alarm or light stays on all the time) or open-circuited (it never comes on, even when it

should). Switch off the ignition, disconnect all wires from the sending unit, and test with an ohmmeter on the R × 1 scale (see *Boatowner's Mechanical and Electrical Manual*) from the sending unit terminal to a good ground. A temperature warning unit should read infinite ohms, unless the engine is overheated, in which case it reads zero. An oil warning unit reads zero with the engine shut down and infinite ohms at normal operating pressures.

Temperature and Oil Pressure Gauges.

Positive current is fed from the battery via the ignition switch to the gauge and from there down to a variable resistor on the engine block and then to ground (see Figure 5-13).

To test a gauge:

1. Test for 12 volts from the gauge positive terminal to a good ground (see Figure 5-14). No volts: the ignition switch circuit is faulty. 12 volts: proceed to the next step.

2. Disconnect the sensing line (that goes to the sending unit) *from the back of the gauge*. A temperature gauge should go to its lowest reading; an oil pressure gauge to its highest reading.

3. Connect a jumper (a screwdriver will do) from the sensing line terminal on the gauge to the negative terminal on the gauge (or a good ground on the engine block if there is no negative terminal). A temperature gauge should go to its highest reading; an oil pressure gauge to its lowest reading.

4. If the gauge passed these tests it is OK. Reconnect the sensing line and disconnect it at the sending unit on the engine. A temperature gauge should go to its lowest reading; an oil pressure gauge to its highest reading. Short the sensing line to the engine block. A temperature gauge should go to its highest reading; an oil pressure gauge to its lowest reading. If not, the sensing line is faulty (shorted or open-circuited).

5. To test a sending unit, switch off the ignition,

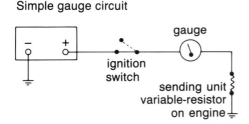

Simple gauge circuit

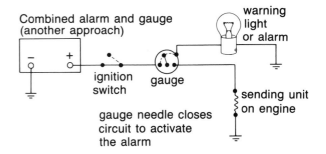

Combined alarm and gauge (another approach)

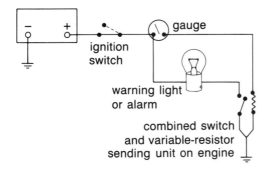

Combined alarm and gauge

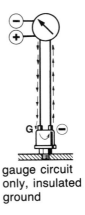

gauge circuit only, insulated ground

Figure 5-13. *Gauge circuits. (Courtesy VDO)*

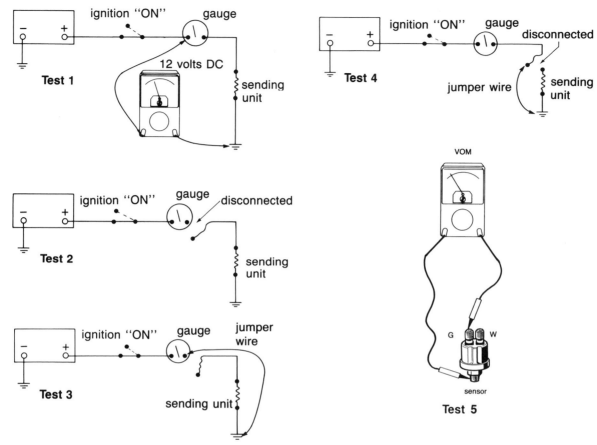

Figure 5-14. *Testing a gauge circuit and sending unit.*

disconnect all wires, and test with an ohm-meter on the R × 1 scale from the sending unit to a good ground. A temperature sender varies from around 700 ohms at low temperatures through 200-300 ohms at around 100 °F (40 °C), down to almost zero ohms at high temperatures: 250 °F (120 °C). An oil pressure sender varies from around zero ohms at no pressure to around 200 ohms at high pressure.

Tank Level Gauges (Electronic). A float is put in the tank either in a tube or on a hinged arm. As the level comes up, the float or arm rises, moving a contact arm on a variable resistance (a kind of rheostat). Positive current is fed from the battery via the ignition switch to the gauge, down to the resistor, and from

there to ground (see Figure 5-15). The gauge registers the changing resistance.

The same tests apply to the gauge, the sensing line, and the sending unit as for an oil pressure gauge. Sending unit resistances are similar—near zero ohms on an empty tank, up to around 200 ohms on a full tank.

Should everything check out OK, but the unit always reads "empty," the float on the sending unit is probably saturated or else there is a mechanical failure—for instance, a broken or jammed arm. A saturated swinging-arm-type float can be made temporarily serviceable by strapping a piece of closed-cell foam to it.

Pneumatic Level Sensors. These are relative newcomers on the market but are likely to become very popular due to their versatility and simplicity. A tube is inserted to the bottom of the tank and con-

Figure 5-15. *Electronic fuel-tank sensor. Because of its construction, the immersion pipe sensor is only suitable for use with fuel tanks. The float unit is mounted on the guide bar and makes contact with the two resistance wires by the contact springs. In this way, the resistance varies in proportion to the level of the liquid, and the variation is displayed at the fuel gauge. The protective tube and the flooder provide excellent damping. Also shown here are the typical differences between the water-tank sensor (left) and the fuel-tank sensor (right). For electrically conductive media, the electrical resistance element must be positioned in the assembly flange. The level of the liquid is then transmitted mechanically up to the element. (Courtesy VDO)*

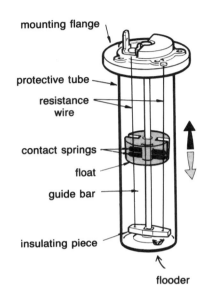

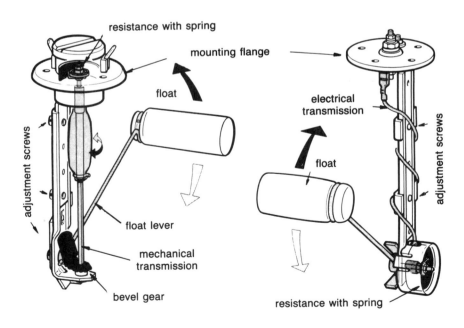

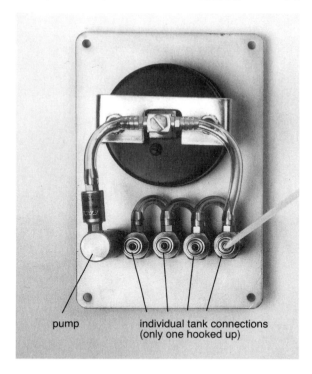

pump   individual tank connections
(only one hooked up)

Figure 5-16. *Pneumatic tank-level sensor. (Courtesy Hart Systems Inc.)*

nected to a small hand pump mounted on the tank gauge panel. The gauge itself is Teed into the tube just below the pump (see Figure 5-16).

When the pump is operated, air is forced down the tube and bubbles out into the tank. Depending on the level in the tank (and hence in the tube) more or less pressure is needed to drive all the fluid out of the tube. The gauge registers this pressure in *inches of water* (in the tank) or *inches of diesel*. A table is drawn up converting the gauge readings to gallons—the conversion will vary from tank to tank depending on the tank size and shape. The gauges come with instructions on how to draw up this table. If the instructions are missing, or with odd-shaped tanks, the best bet is to simply empty the tank, then to keep adding known quantities of fluid (e.g., five gallons at a time), stroking the hand pump and noting the gauge reading after each addition.

Pneumatic level gauges are more accurate than electronic gauges. What is more, apart from leaking connections or kinked tubes, there is nothing to go wrong. As many as 10 tanks can be measured with one gauge by simply switching the gauge and pump into the individual tank tubes. Since there is no fluid in the tubes beyond the level of the tanks themselves, there is no possibility of cross-contamination from one tank to another. This means diesel and water tanks both can be measured with the same unit.

Chapter 6

Maintenance and Repair Procedures, Part One

Before tearing into an engine, take the time to carefully analyze problems. Quite often an over-enthusiastic mechanic will take the wrong thing apart, in the process spending a lot of unneccesary money and destroying the evidence needed to make a correct analysis of the problem. For example, recently we heard of a boatowner who had been experiencing problems with very uneven thrust. A mechanic told him the transmission was defective and changed it. The problem was still there with the new transmission, and the mechanic gave up. It turned out to be nothing more than water in the fuel system!

Remember also that if you pull an engine to pieces but fail to find your problem, you'll be forced to consult a mechanic. The mechanic may have to reassemble the engine and run it in order to find out what the problem was in the first place, then he will have to disassemble it to fix it! This can get very expensive.

The Cooling System

The cooling system normally requires very little attention; the exceptions are the raw-water strainer and sacrificial zincs (see Figure 6-1). Check the strainer at regular intervals as part of routine maintenance—perhaps whenever you change the oil. If the engine shows signs of overheating, first suspect a plugged raw-water strainer.

Sacrificial Zincs. These are essential in a raw-water circuit to protect heat exchangers and oil coolers from galvanic corrosion. A number of poorly-made heat exchangers and oil coolers use brass, or other alloys of zinc, in their construction. If they aren't adequately protected with sacrificial zincs, in the presence of heat, salt water, and the dissimilar metals used in their construction, these units soon start to lose the zinc in their alloys (known as *dezincification*). Dezincification leads to pin-holes in cooling tubes and around welds and solder joints.

Failed heat exchangers and oil coolers can result in expensive engine damage, and can be hard to replace, especially when you're cruising in remote areas. When cruising, inspect zincs every 30 days until you get some idea of the rate at which they are being consumed. Carry plenty of spares, and replace them when they're only partially used up (no more than 50%). Don't wait until they are completely eaten away because by this time galvanic corrosion will already be attacking your engine (see Figure 6-2).

In the absence of replacement zincs, take the threaded cap from an old zinc, place it over the galley stove until the solder in the base of the cap begins to puddle, then jam any available piece of zinc, suitably cut to size, into the solder.

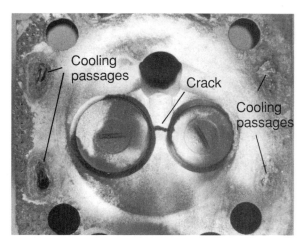

Figure 6-1. *Cracked cylinder head from a raw-water-cooled engine. The cooling passages were badly corroded from failure to change the sacrificial zinc anode. The engine then overheated and cracked the cylinder head.*

Figure 6-2. *A zinc anode in the cooling system of a Yamaha diesel engine.*

Heat Exchangers and Oil Coolers.

In time, corrosion, silt, and sand may block the cooling tubes. The engine will show a slow but steady rise in temperature. Many heat exchangers and oil coolers have removable end caps, which allow the tubes to be cleaned out. You may need to push a rod through the tubes in order to clear blockages. In this case, use a wooden dowel (not metal), and do not use too much force—the relatively soft tubes can be easily dam-

aged. For measures to deal with a failed heat exchanger, see page 82.

Flushing Raw-water-cooled Engines.

Silt and other loose deposits can sometimes be flushed out of the block of a raw-water-cooled engine by the following: remove some item (e.g., a thermostat) that gives access to the cooling passages, open the cylinder block drain valve (or better still, remove it to create a larger hole), and run water from a hose through the block. Use as much pressure as possible.

Scale and salt deposits can generally only be removed chemically. Drew Ameroid Marine (One Drew Plaza, Boonton, NJ 07005; 201-263-7600) is one company with world-wide distributors that manufactures solutions for this purpose. An engine with a *cast-iron block* and *cast-iron cylinder head* can be descaled with their Saf Acid or, in a pinch, with phosphoric or oxalic acid. (The latter is used in a solution of 8 ounces to 2.5 gallons of water. You can get it from engine-supply houses and in boatyards that use it for cleaning hulls.) The acid solution is pumped around and around the engine, dissolving the scale. When all the scale is dissolved, *thoroughly flush the engine with fresh water*, neutralize any remaining acid with an alkaline solution (2 ounces of sodium carbonate dissolved in 2.5 gallons of water is recommended), then thoroughly flush again. *Do not use this procedure on aluminum blocks or cylinder heads without first seeking professional advice.*

Special commercial chlorinated solvents are sometimes used on aluminum (or Drew Ameroid's Ameroid OSC—one-step cleaner). Once again, seek professional advice—it is quite possible to do expensive damage. I once worked on a 2,000 h.p. engine with a salt problem. A solution used to dissolve the salt was not properly flushed out. The cooling system was then filled with an antifreeze solution. The two chemicals reacted to precipitate out an insoluble goo that completely plugged the entire engine! Two months, and several chemical companies later, no one had been able to come up with a chemical that would dissolve the goo without also eating away some part of the engine.

Freshwater Circuits.

Antifreeze should be added to freshwater circuits whether the boat is in a cold climate or not. The antifreeze contains various rust and corrosion inhibitors, which help to keep the cooling system clean and undamaged. Since these inhibitors get used up over time, you should change the antifreeze annually in order to renew them, or at the first signs of rust in the cooling system.

Bleeding a Cooling System. When you start an engine after any part of the cooling system has been drained, you must check the flow of coolant, particularly on the raw-water side. If a water pump is airbound, or water fails to circulate for any other reason, the vanes will strip off rubber impellers in seconds; cylinder heads can crack in minutes.

Bleed an air-bound cooling system just as you would a fuel system, starting from the water source and working toward the pump. Once the pump is primed, it should push the coolant through the rest of the system. However, on occasion air-locks can form in hoses, stalling out the water flow—the long hose runs associated with the heat exchangers in domestic water heaters are particularly prone to this. If this is a problem, progressively loosen hose connections around the system, allowing water to vent out of each connection before retightening.

Hoses and Hose Connections. Suction-side hoses to water pumps need to be of the non-collapsing type, preferably with an internal *stainless steel* spring. In time, hoses that are subjected to heat soften and bulge, or crack, and need replacing.

Hose clamps must be of stainless steel, including the worm screw. All hose connections below the waterline need to be double-clamped. Once a year, you should loosen hose clamps and inspect the band where it has been *inside the screw housing*—the bands are prone to failure from crevice corrosion at this point.

Thermostats. Thermostats are generally very reliable, but occasionally do fail, normally in the closed position. Failure leads to overheating. The thermostat housing is almost always near the top of the engine, at the front (the crankshaft pulley end), with one or more cooling hoses coming out of it (see Figure 6-3 and Figure 6-4). Generally, it is held down with a couple of bolts. Once these are removed, gently pry up the housing. The thermostat can be lifted out and tested as described on page 72. Thermostats are rela-

Figure 6-3. *Thermostat housing on a Volvo MD17C.*

Figure 6-4. *Thermostat removed from a Volvo MD17C.*

tively cheap, so if you go to the trouble of taking yours out, you might as well replace it.

Water Pumps. The pumps are generally of the piston, diaphragm, or flexible-impeller type on raw-water circuits, and of the centrifugal type (standard on automobiles—see Figure 6-5) on freshwater circuits. As long as you keep the drive belt tight, the latter rarely give problems, but if yours does, simply exchange it for a new one. Signs that a replacement is needed are leaking seals around the drive shaft, leaks from the drain hole on the underside of the pump body, squealing bearings, undue play in the bearings, or any roughness in the bearings when you turn the pump by hand with the belt removed. *Note: if the pump is ever run dry, the seals will fail rapidly.*

Piston-type pumps suck water into a cylinder through an inlet valve and push it back out through a discharge valve on the next stroke. The piston is generally sealed in its cylinder with a rubber O-ring, or on some older engines with a dished leather washer bolted to the bottom of the piston.

The piston seal and valves are the only items likely to cause trouble. The valves and their seats slowly deteriorate due to corrosion, leading to a gradual loss of pumping capability. Trash stuck in the valve seats can completely disable the pump. (A raw-water circuit must have an effective filter.)

Over a winter's lay-up when the engine is drained and left for several months, the piston seal will sometimes stick to its cylinder wall and then tear when the engine is first started in the spring. The puzzled owner knows that everything was functioning just fine when the boat was laid up, and can't make out why there is no water flow in the spring. Always carry a spare O-ring or leather on board. In an emergency, any thin piece of leather (e.g., from an old wallet), cut in a circle and worked up the sides of the piston with a generous smearing of grease, will temporarily replace a leather washer.

Diaphragm-type water pumps operate in just the same way that diaphragm-type lift pumps do. Valves slowly corrode or collect trash. The only other likely point of failure is the diaphragm. The pumps generally have a hole in their base: if the diaphragm fails, water will dribble out. Diaphragms are easily replaced by removing the pump cover.

Figure 6-5. *Cutaway of an engine-mounted centrifugal water pump. (Courtesy Caterpillar Tractor Co.)*

Flexible Impeller Pumps

Flexible-impeller pumps form the vast majority of raw-water pumps in small boats. Since they are a relatively common source of problems, they need to be looked at in some detail (see Figure 6-6 and Figure 6-7).

The principal causes of failure are running dry and swollen impellers caused by chemicals (generally oil and diesel) in the raw water. In the former case, the impeller vanes are likely to strip off; in the latter, the impeller jams in the pump's body. In either case, a new impeller is needed. Always carry a couple of spares on board. The original pump will almost certainly have a *neoprene* impeller, but when you buy replacements, get *nitrile* or *viton* impellers (preferably the latter) because they have a far greater resistance to chemicals. They are, however, a little more expensive than neoprene impellers.

Flexible-impeller pumps like to be used often. If left for long periods without running (e.g., over the winter), the vanes tend to stick to the pump housing. When the pump is restarted, it may blow fuses (if electrically driven) or strip off its vanes. If this is a constant problem, try fitting an overthick cover gasket (see "Reassembly" below—this may result in a small loss of pumping capacity). Otherwise, after a period of shutdown, the pump cover will need loosening until

Figure 6-6. *A raw-water pump from a Volvo MD17C.*

groove of the two pulleys; any misalignment will be clearly visible. Pumps which operate in sandy water will experience a gradual loss of performance due to wear on the pump housing, cover plate, wear plate (if fitted), and impeller.

Disassembly. Despite the variety of flexible impeller pumps in service, most share the same construction details (see Figure 6-8A and Figure 6-8B). Removal of the end cover (four to six screws) will expose the impeller. Almost all impellers are a sliding fit on the drive shaft, either with splines, square keys, Woodruff keys, one or two flats on the shaft, or a slotted shaft. Using a pair of needle-nose pliers, grip the impeller and pull it out. Some impellers are sealed to their shafts with O-rings; most are not. If the impeller

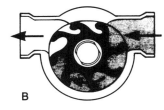

A   B   C

Figure 6-7. *Operation of a variable-volume pump with a flexible impeller. The housing is flattened on the top, which pushes the impeller's blades toward the shaft. As the blades pop up (A), the action creates a vacuum, which draws liquid into the pump. Each blade draws in liquid and carries it to the outlet port (B). When the blades compress, they expel the remainder of the liquid (C). The flow is continuous and uniform. (Courtesy ITT/Jabsco)*

won't come out, it may be one of the few locked in place with a set screw or allen screw (some Volvo and Atomic Four engines in particular). On some of these, you can pull out the impeller far enough ($1/2''$) to release the screw, but if the screw is inaccessible, the drive side of the impeller will have to be disassembled and the impeller knocked out on its shaft.

Inspect the impeller. The vanes should have rounded tips (not worn flat) and show no signs of swelling, distortion, or cracking. If you're in any doubt, replace the impeller. If an O-ring is fitted to the shaft, check it for damage. If the impeller has a tapered metal sleeve on its inner end (*extended insert* impellers), inspect the sleeve and discard the impeller if there is any sign of a step where the sleeve slides into the shaft seal.

If it is necessary to remove the cam (e.g., to replace a wear plate, see below), loosen the cam retaining screw, tap the screw until the cam breaks loose, then remove the screw and cam. Remove sealing compound from all surfaces.

the pump starts spinning and the impeller is thoroughly lubricated with water.

Other common problems are leaking seals (quite likely as a result of running dry) and worn or corroded bearings. Apart from normal wear, bearings will be damaged by water from leaking seals, an overtightened belt, and misalignment of drive pulleys or couplings. Optimum belt tension will allow the longest stretch of belt to be depressed by $1/3''$ to $1/2''$ with moderate finger pressure. Pulley alignment can be checked by removing the belt and placing a rod in the

---

Troubleshooting Chart 6-1.
Flexible Impeller, Vane, and Rotary Pump Problems: No Flow.

If the pump is belt-driven, is the pump pulley turning? **YES**	**NO**	Tighten or replace the drive belt.
If a clutch is fitted, is it working? (The center of the pulley will be turning with the pulley itself.) **YES**	**NO**	Adjust or replace the clutch.

For gear-driven pumps and for belt-driven pumps on which the pulley is turning, proceed as follows:

Remove the pump cover: Are the impeller vanes intact? **YES**	**NO**	Replace the impeller and track down any missing vanes. The pump probably ran dry; find out why. Check for a closed seacock, plugged filter, collapsed suction hose, or excessive heeling that causes the suction line on a raw-water pump to come out of the water. Less likely is a blockage on the discharge side causing the pump to overload.
Are the vanes making good contact with the pump body? **YES**	**NO**	The impeller is badly worn and needs replacing.
Does the impeller turn when the pump drive gear or pulley turns? **YES**	**NO**	The impeller, drive gear, or pulley is slipping on its shaft, or the clutch (if fitted) is inoperative. Repair as necessary.

If the impeller turns, the pump may just need priming. Otherwise the suction or discharge line must be blocked. Check for a closed seacock, plugged filter, collapsed suction hose, or excessive heeling that causes the suction line on a raw-water pump to come out of the water.

Some impellers have a *wear plate* at the back of the pump chamber. If fitted, the plate can be hooked out with a piece of bent wire. Note the notch in its top; this aligns with a dowel in the pump body. If the wear plate is grooved or scored, replace it.

Shaft seals are of three types:

1. *Lip-type* seals which press into the pump housing and have a rubber lip that grips the shaft.
2. *Carbon/ceramic* seals, in which a ceramic disc with a smooth face seats in a rubber "boot" in the pump housing while a spring-loaded carbon disc, also with a smooth face, is sealed to the pump shaft with a rubber sleeve or O-ring. The spring holds the carbon disc against the ceramic disc, and the extremely smooth faces of the two provide a seal.
3. An external *packing gland* (stuffing box), which is the same as a propeller shaft packing

Figure 6-8B. *An exploded view of the pump shown in Figure 6-8A. 1. Cover. 2. Gasket. 3. Impeller (splined type). 4.Wear plate. 5. Cover retaining screws. 6. Cam (mounted inside the body of the pump). 7. Cam's retaining screw. 8.Pump body. 9. Slinger (to deflect leaks away from the bearing). 10. Bearing. 11. Bearing's retaining circlip. 12. Shaft retaining circlip. 13. Outer seal. 14-16. Inner seal assembly, lip type. 17. (alternative) Inner seal assembly, carbon-ceramic type. 18. Pump shaft. 19. Drive key. (Courtesy ITT/ JABSCO)*

Figure 6-8A. *A typical flexible-impeller pump.*

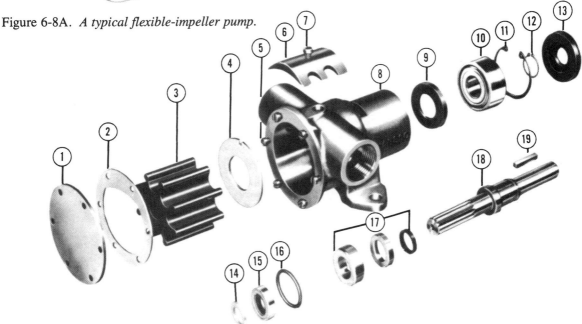

gland. These are not very common. (For care and maintenance, see page 178.)

Although it's possible to hook out and replace some carbon/ceramic seals with the shaft still in place (this cannot be done with lip-type seals), in most cases the shaft must be taken out. To do this, take apart the drive end of the shaft. First unbolt the pump from its engine or remove the drive pulley if belt driven; then proceed as follows:

If the pump has a seal at the drive end of the shaft (part number 13 in Figure 6-8B), hook it out. When removing any seals, be extremely careful not to scratch the seal seat. Behind the seal in the body of the pump, you will usually find a bearing-retaining circlip. (Pumps bolted directly to an engine housing may not have one.) Remove the circlip, flexing it the minimum amount necessary to get it out. If it gets bent, replace it rather than straightening it out.

Support the pump body on a couple of blocks of wood and tap out the shaft, hitting it on its *impeller* end (see Figure 6-9). (*Exception: pumps with impellers fastened to the shaft—these must be driven out the other way.*) Do not hit the shaft hard; be especially careful not to burr or flatten the end of the shaft. It is best to use a block of wood between the hammer and the shaft, rather than to hit the shaft directly.

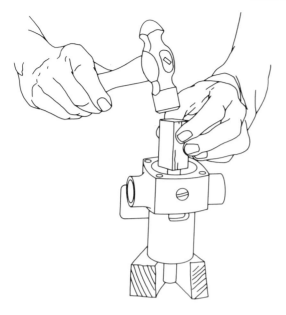

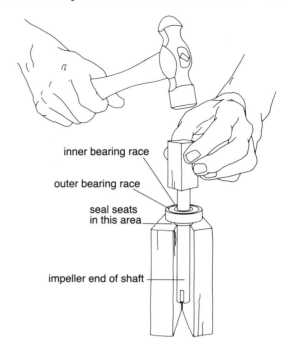

inner bearing race

outer bearing race

seal seats
in this area

impeller end of shaft

Figure 6-9. *Removing a pump shaft. Remove the end cover and impeller. Support the pump body with a couple of blocks of wood, impeller end up. Protect the end of the impeller shaft with a block of wood and lightly tap the shaft free.*

Figure 6-10. *Removing a bearing from the pump shaft. Remove the retaining circlip from the shaft; support the bearing with a couple of wooden blocks placed under the inner bearing race; protect the shaft end with a wooden block and lightly tap the shaft out of the bearing.*

If the shaft won't move, take another look for a bearing-retaining circlip. If there truly isn't one, try hitting a little harder. If the shaft remains stuck fast, try heating the pump body in the area of the bearing with hot water or gentle use of a propane torch. The shaft will come out complete with bearings.

The main shaft seal can now be picked out from the impeller side of the body with a piece of bent wire. There may be another bearing seal on the drive side, and quite probably a slinger washer between the two. If the washer drops down inside the body, retrieve it through the drain slot.

To remove the bearings from the shaft, take the small bearing-retaining circlip off the shaft, support the bearing with a couple of blocks of wood placed *under the inner bearing race*, and tap out the shaft, hitting it on its drive end (see Figure 6-10) using a block of wood between the hammer and the shaft once again.

Inspect the shaft for signs of wear, especially in the area of the shaft seal. Spin the bearings and discard them if they are rough, uneven, or if the outer race is loose. Scrupulously clean the pump body, paying special attention to all seal and bearing seats. *Do not scratch bearing, seal, or seating surfaces.*

Reassembly. To fit new bearings to a shaft, support the inner race of the bearing and tap the shaft home. To make the job easy, first heat the bearing (e.g., in an oven to around 200 °F [93 °C], but no more) and cool the shaft in the icebox. The shaft should just about drop into place. Replace the bearing-retaining circlip, flat side toward the bearing.

Now is the time to put the new shaft seal in the pump body—also the inner bearing seal (if fitted). Lip-type seals have the lip *toward the impeller*. Carbon/ceramic seals have the ceramic part in the pump body, set in its rubber boot with the shiny surface facing the impeller.

Lip-type seals are lightly greased (petroleum jelly or a Teflon-based, waterproof grease), *but carbon/ceramic seals must not be greased*. On the latter, the seal faces must be wiped spotlessly clean—even finger grease must be kept off them—and the seals lubricated with water.

All seals must be centered squarely and pushed in evenly (see Figure 6-11). If the seal is bent, distorted, or cockeyed in any way, it is certain to leak. A piece of hardwood dowel the same diameter as the seal—or a little bigger—makes a good drift. The seal is pushed down until it is flush with the pump chamber.

If the pump has a slinger washer, slide it up through the body drain and maneuver the shaft in from the drive side of the body, easing it through the slinger and into the shaft seal (see Figure 6-12). Pass the shaft very carefully through any seals.

Seat the bearings squarely in the pump housing, support the pump body, and evenly drive the bearings home, applying pressure to the *outer* race (a socket with just a little smaller diameter than the bearing works well—see Figure 6-13). Heating the pump body (but not with a propane torch this time because you may damage the seals) and cooling bearings will help tremendously. Refit the bearing-retaining circlip with the flat side to the bearing, and press home the outer bearing seal (if fitted), lip side toward the pump impeller.

Turn now to the pump end (refer to Figure 6-8B). If the pump has a carbon/ceramic seal (see Figure 6-14), clean the seal face, lubricate it with water, and slide the carbon part up the pump shaft, with the smooth face toward the ceramic seat. Some seals use *wave* washers to maintain tension between the carbon seal and the seat; most use springs. (If the new seal has both, discard the wave washer.) Replace the wear plate, locating its notch on the dowel pin. Lightly apply some sealing compound (e.g., Permatex) to the back of the cam and to its retaining screw. *Loosely* fit the cam. Lightly lubricate the impeller with dishwashing liquid and push it home, bending down the vanes in the opposite direction to pump rotation. Replace the gasket and pump cover. Tighten the cam screw.

The correct gasket is important—too thin, and the

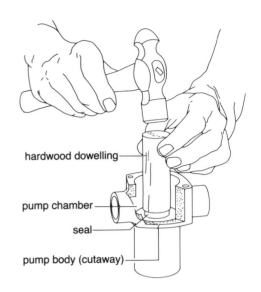

hardwood dowelling

pump chamber

seal

pump body (cutaway)

Figure 6-11. *Replacing a pump seal. Seat the seal squarely in its housing. Place a length of hardwood doweling (the same or slightly larger in diameter than the seal) on the seal. Make sure that the doweling seats on the metal rim of the seal, not on the rubber lip. Using a hammer, tap in the seal very softly until it is flush with the body of the pump.*

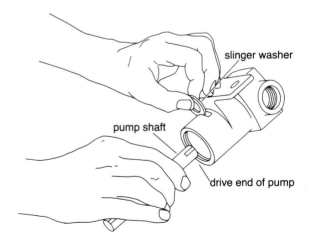

slinger washer

pump shaft

drive end of pump

Figure 6-12. *Replacing a slinger washer. Push the washer up through its slot in the pump body, threading the shaft through it and into the seal.*

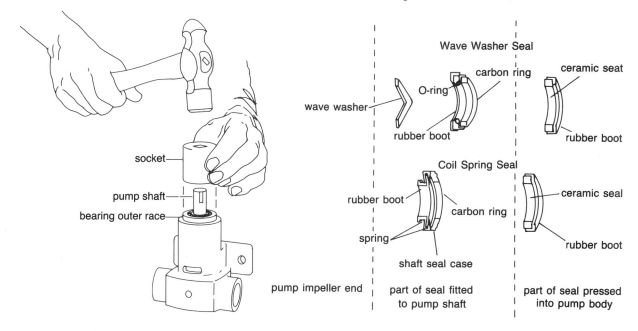

Figure 6-13. *Driving home a bearing. Use a ratchet-drive socket that has a diameter slightly smaller than the outer race of the bearing. Be sure the bearing is squarely seated in the housing before you drive it home.*

Figure 6-14. *Carbon-ceramic seals are of two basic types: wave-washer (uncommon) and coil-spring. In either case, the ceramic seal and rubber boot are pressed into place in the pump body, and the carbon-ring seal and retainer are fitted to the shaft. (Courtesy ITT/Jabsco)*

impeller will bind; too thick, and pumping efficiency is lost. Most pump gaskets are 0.010″ (ten thousandths of an inch) thick, but on larger pumps this may be 0.015″. As noted previously, some impellers on pumps used only intermittently have a tendency to stick in their housings during periods of shutdown. To stop this, loosen the pump cover on initial start-up, then tighten it. You can achieve much the same result with a small loss in pumping efficiency, without loosening the screws, if you fit an overthick gasket.

When refitting a flanged pump to an engine housing, be sure the slot in the pump drive shaft, or the drive gear, correctly engages the tang or gear on the engine, and make sure the pump flange seats squarely *without pressure*. Pulley-driven pumps must be properly aligned with their drive pulleys, and the belt correctly tensioned.

The Exhaust System

Exhausts on many auxiliary sailboats tend to slowly plug up with carbon, due to prolonged hours of low-load operation when running a refrigerator or charging batteries at anchor. An exhaust should be broken loose and inspected annually. A heavily sooted exhaust will need cleaning, as will the exhaust passages on the engine and probably the turbocharger (if fitted).

Water-cooled exhausts corrode because of exposure to the combination of hot exhaust gases and salt water. Downstream from where the water enters the system, good-quality, wire-reinforced steam hose is the best material for exhaust pipes. Fire-retardant fiberglass is best for water-lift mufflers. If the raw-water circuit fails, both materials are likely to burn up, although any high-temperature engine alarm (if fitted) should sound before this happens. All hose joints should be double clamped with all-stainless hose clamps. Inspect the hoses and clamps annually (see above).

Exhaust Injection Elbow. Sooner or later the injection elbow will corrode through. The elbow is frequently custom-made and hard to replace, so you should carefully inspect it annually and carry a spare on board. If you have to have a new one custom-fabricated, *make sure the welder uses similar welding rods to the metal used in fabricating the elbow* (e.g., stainless rods on stainless steel). Dissimilar metals will greatly accelerate the rate of corrosion. Under no circumstances have the joint brazed.

You can temporarily repair a holed elbow by tightly wrapping a strip of rubber cut from an inner tube around the elbow and clamping with two hose clamps (see Figure 6-15). Alternatively, use one of the underwater epoxies (Coppers and Co., Z-Spar, Marine-Tex) to patch the hole.

Siphon Breaks. All proper water-cooled exhaust installations require one or more siphon breaks (see Chapter 9). These, however, tend to plug with salt crystals, which can render them ineffective, allowing them at the same time to spray salt water over the engine and its electrical systems. The valve should be periodically removed and rinsed in warm fresh water. Better yet, remove the valve element (or fit a tee in place of the siphon break in the first place), add a hose to the top, and vent this well above the waterline (into the cockpit works well—see Figure 6-16).

Cleaning Turbocharger Wheels. Turbochargers (see Figure 6-17), as a rule, should be left to the experts. However, if removal of the inlet and exhaust ducting and the tests outlined on page 84 reveal no problems other than a dirty compressor wheel or turbine, you can generally clean these without help.

Mark both housings and the center unit with scribed lines so that you can put them together in the same relationship to one another. Allow the unit to cool before removing any fasteners or it may warp. If the housings are held together with large snap rings (circlips), leave them alone—they will come apart easily enough but may require a hydraulic press to put

Figure 6-15. *A corroded galvanized exhaust elbow. The hot gases and water from the exhaust have eaten through it. The same elbow (below) patched with an inner tube and hose clamps. The repair lasted for 200 hours of engine running time.*

together! Those held with bolts and large clamps can be taken apart.

If the housings are difficult to break loose, tap them with a soft-faced hammer or mallet. Pull them off squarely to avoid bending compressor wheel or turbine blades (see Figure 6-18). The wheel and turbine can be cleaned with non-caustic solutions only (degreasers work well) using *soft-bristle* brushes and plastic scrapers. Do not use abrasives; resultant damage to the blades will upset the critical balance of the wheel and turbine. Make no attempt to straighten bent blades—if you find any, the turbocharger demands a specialist's help.

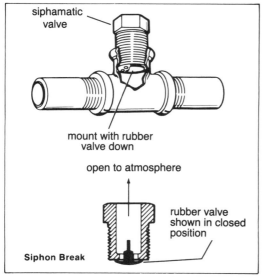

(Courtesy Kohler)

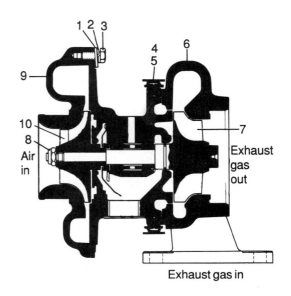

Exhaust gas in

1. Washer
2. Lock washer
3. Bolts fastening compressor cover
4. V-clamp
5. V-clamp lock nut
6. Turbine housing
7. Exhaust turbine
8. Main shaft nut
9. Compressor housing
10. Compressor turbine

Figure 6-17. *Holset 3LD/3LE Turbocharger. (Courtesy Holset Engineering Co. Ltd.)*

Figure 6-16. *A vented loop on an engine cooling circuit. The hose at the top discharges into the cockpit so that cooling water won't spray over the engine if the vent fails in the open position.*

Dismantling the center unit is not recommended. If you do attempt it for any reason, note that the center retaining nut has a *left-handed* thread.

The procedure for reassembly is the reverse of disassembly. Be sure to line up accurately all scribed marks. When you are done, spin the unit by hand to make sure it turns freely. *Make sure that the air-inlet passages, air filter, and internal engine air passages are scrupulously clean.* Before starting the engine, crank it for a while with the throttle closed to get the oil up to the turbocharger bearings.

Figure 6-18. *Removing a turbocharger housing. (Courtesy Perkins Engines Ltd.)*

Governors

Problems with governors are usually indicated by the engine's failing to hold a set speed, particularly on start-up. *Hunting* is the most common symptom—the engine continuously and rhythmically cycles up and down. However, not all such speed fluctuations are caused by the governor, so first check for excessive load changes (e.g., a microwave on less than full power is actually going from full power to no power at timed intervals; this will cause the governor on a small diesel generator to work hard), misfiring cylinders (see page 77), and poor injection (see later in this chapter).

If these checks fail to reveal a problem, most likely some part of the governor mechanism or fuel injection pump control linkage is sticking. The governor has to overreact to any change in load to overcome resistance to movement of the fuel rod. Then the rod moves with a jerk and goes too far, causing the engine to overspeed or underspeed. The governor then overreacts once again, but in the other direction, and the engine underspeeds or overspeeds, and so on. Excessive slack or play in the fuel pump control linkage (throttle linkage) will cause similar behavior.

Many governors are inside the fuel injection pump housing. Injection pumps are items for the experts—if the problem cannot be solved by cleaning and adjusting the external fuel control linkage, call in a professional.

If the governor is inside the engine (as opposed to

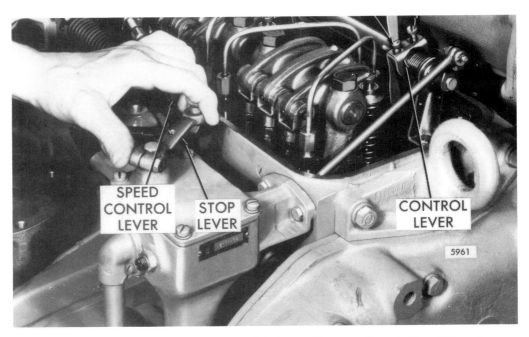

Figure 6-19. *Detroit Diesel fuel-rack control linkage. (Courtesy Detroit Diesel Corp.)*

the fuel injection pump), the governor mechanism itself may be causing the problem. In this case, quite likely a build-up of sludge from a failure to carry out adequate engine oil changes, or from prolonged low-load operation, is interfering with correct governor operation. You will have to clean the governor.

Detroit Diesels generally have an externally mounted governor operating an *injector control tube* (see Figure 6-19). If you remove the valve cover, all the operating levers will be accessible. You should be able to move the injector control tube, and thus all the injector racks, through their full range without any rough spots or sticking. If friction exists, disconnect the injectors one by one and test again. If the friction ceases, check the individual injector rack. If it is sticking, it may simply be because the injector hold-down clamp is improperly positioned or too tight (try loosening the clamp, repositioning it, and tightening), but more likely the injector needs an overhaul. If there is still friction in the linkage after all the injectors are disconnected, check the various support brackets on the control tube for binding, and the position of the control lever (running to the governor) on the control tube.

The only other likely maintenance item on a governor is an occasional adjustment of the idle setting on the speeder spring. There is normally no cause to alter this. If the engine will not idle correctly, it's almost certainly due to some other problem. The idle setting should only be adjusted when the engine is running well. Engine-mounted governors have a screw and locknut somewhere on the outside of the block. Governors inside injection pumps generally have an external low-speed screw acting directly on the throttle control lever.

Somewhere there will also be a maximum-fuel setting screw and lock-nut that will almost certainly be tied off with lockwire and sealed (see Figure 6-20). *Do not tamper with it.* Breaking the seal automatically voids any engine warranty. If it's not sealed and if the engine appears to be overloaded at full throttle (making black smoke, overheating), the maximum fuel setting should be reduced. *Increasing maximum fuel settings above manufacturer's set points can lead to engine seizure.*

Fuel Injection Pumps

Before jumping to the conclusion that you have a defective fuel injection system, always check every other likely cause of your problem, particularly an obstructed airflow through the engine (plugged air filters, collapsed ducting, poor turbocharger or blower performance, a carboned-up exhaust pipe, etc.).

The only user-serviceable parts of a fuel injection system are the fuel strainer and rubber diaphragm in the lift pump, and the diaphragm found in some fuel injection pumps. Lift pump servicing has already been covered (see page 64).

Diaphragm Fuel Injection Pumps. A few engines have vacuum-type governors (see page 22). In this case, the injection pump has a diaphragm in the

Governor control lever (throttle)   Idle speed fuel setting screw

Maximum fuel screw with lockwire and seal   Throttle (governor control) linkage

Figure 6-20. *Maximum fuel setting screw on a Volvo MD17C.*

back of it. This diaphragm pushes against the fuel-control rack on one side and is controlled by a vacuum line to the air inlet manifold on the other side. A holed diaphragm can lead to a loss of power, excessive black smoke, a very rough idle, and overspeeding.

Test an injection pump diaphragm with the engine off. The diaphragm is contained in a round housing at the back of the pump (the end not attached to the engine). Coming out of the top of this housing is a vacuum-sensing line that leads to the air-inlet manifold. Disconnect this line at the pump.

At the opposite end of the pump, just above the flange that holds it to the engine, will be a protective cap, inside of which is the fuel-control rack. Remove this cap. On the side of the pump, below and just forward of the vacuum line, is the pump-control lever (throttle). Hold this in the stop position.

If you place a finger tightly over the vacuum connection on the pump and let go of the control-rod lever, the fuel-control rack at the engine end of the pump should move a short distance then stop, as long as the vacuum fitting is kept blocked by your finger. If there are any leaks in the diaphragm, or around the seal to its housing, the control rack will keep moving (perhaps slowly) until it is as far out as it can go. In this case, the diaphragm needs inspecting, and probably replacing.

You can get to the diaphragm by removing its cover (four screws) and undoing the bolt holding it to the fuel rack.

Injection Pump Oil Reservoirs. Although most pumps are lubricated with diesel fuel, some in-line jerk pumps have an oil sump that contains regular engine oil. This becomes diluted with diesel over time and occasionally needs changing. The sump will have an oil drain plug and a dipstick or level plug. In the latter case, take out the level plug and fill the sump with oil until it runs out of the hole. Replace the plug. *No other area of the fuel pump is serviceable in the field, and it should be left strictly alone.*

Injectors

Lucas CAV recommends: "In the absence of specific data, a figure of 900 hours of operating between servicing is a useful guide for the boatowner." Aside

from pulling the injectors to have them serviced, injectors should be left alone. Individual injector needle valves are matched to their bodies to within 0.00004″ (four one-hundred thousandths of an inch). Equipment of this degree of precision needs to be disassembled by specialists. "It is not possible for the owner or crew to recondition or service an injector without the essential nozzle setting outfit, special tools, technical data, and service training," states the CAV handbook. "Any tampering or attempts at servicing without these essentials will always make matters worse." The following information is therefore given only for those in a dire emergency, and after all other procedures have failed to solve a problem.

Servicing of a fuel injection system requires *extreme cleanliness*. Before you attempt to break loose fuel lines or remove injectors, the area around them should be thoroughly cleaned. The instant any fuel lines are disconnected both loose connections must be capped. Once an injector is removed from a cylinder head, plug its hole to prevent dirt from falling into the cylinder.

Injectors are held in place by a metal plate bolted to the cylinder head, or they are screwed directly into the head (see Figure 6-21). The former can sometimes be hard to break loose—dribble a little penetrating oil down the side of the injector an hour or two before you pull it out.

On some engines (for example, the Volvo Penta series 2000 and Detroit Diesel series 71), the injector is in a sleeve that is directly cooled by the engine's cooling circuit. Occasionally the sleeve sticks to the injector and comes out with it. The block should therefore be drained of coolant before attempting to pull any injectors. This eliminates the risk of coolant running into a cylinder. Once again, dribble penetrating oil down the sides of the injector, let it sit for a while, then loosen the injector by working it from side to side. If the sleeve comes out, you will have to call in a trained mechanic because installation of a new sleeve requires special tools.

Injectors, and many fuel lines, are sealed with copper washers. Do not lose any of these, and be sure that they go back in the right place on reassembly.

Testing an Injector. Now that the injector is out of the cylinder head, you can check its operation by reconnecting it to its delivery pipe, bleeding the line,

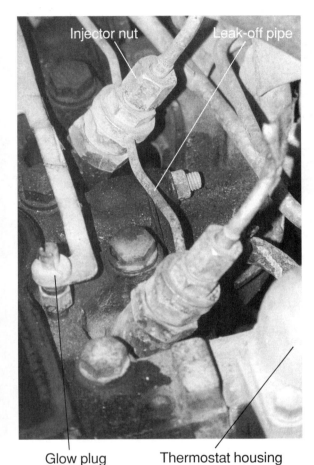

Figure 6-21. *Screw-in-type injectors.*

Figure 6-22. *Proper injector spray patterns.*

opening the throttle, cranking the engine over, and observing the spray pattern (see Figure 6-22). Do this *only* after you have loosened the delivery pipes to the other injectors to prevent the engine from starting. The engine should be rotated at a speed of at least 60 rpm (normal cranking speed on most small diesels is more than adequate). Each type of injector will produce a distinct spray pattern, but all should have certain features in common:

1. High degree of atomization of the diesel;
2. Strong, straight-line projection of the spray from the nozzle as a fine mist with no visible streaks of unvaporized fuel;
3. No dribbling or drops of fuel. (The nozzle should remain dry after injection is complete.);
4. Fuel should come out of all the holes in the nozzle in even proportions.

Figure 6-23 illustrates different spray patterns. If you have an injector that you know is good, test this one first to familiarize yourself with a correct spray pattern before going on to test suspect injectors. If you

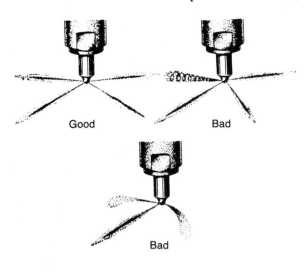

Figure 6-23. *Pressure-testing the injectors. Check the spray pattern and also that the fuel jets cease simultaneously at all four holes and that they do not drip afterwards. (Courtesy Volvo Penta)*

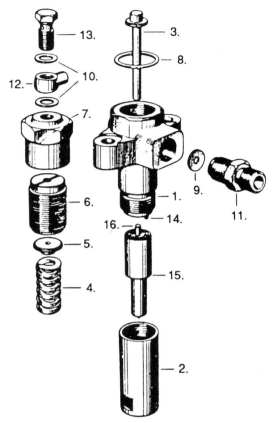

1. Nozzle holder	9. Joint washer
2. Nozzle nut	10. Joint washers
3. Spindle	11. Inlet adaptor
4. Spring	12. Leak-off connection
5. Upper spring plate	13. Banjo screw
6. Spring cap nut	14. Dowel
7. Cap nut	15. Nozzle
8. Joint washer	16. Needle valve

Figure 6-24. *Exploded view of a multi-hole injector. (Courtesy Lucas CAV Ltd)*

are in doubt about the spray patterns, take the injectors to a fuel injection shop to have them tested. Testing (though not re-building!) is cheap.

Whenever an injector is tested, you must keep well out of the way. The diesel fuel is fine enough, and has more than enough force, to penetrate skin and blood vessels. It can cause severe blisters and blood poisoning.

If the spray pattern is defective or the nozzle drips, the injector nozzle needs cleaning. Cleaning the carbon off the outside of the nozzle and using a very fine wire to clear the holes in the nozzle may be sufficient. A brush with *brass* bristles and diesel fuel should be used on the nozzle—steel brushes should never be used because they can damage the holes. If a proper set of injector-cleaning prickers is not available, a strand of copper wire may work. The pricker off a primus stove or kerosene lantern is another alternative. Do not enlarge the holes or break off a pricker in an injector hole.

Disassembly. If you have to disassemble an injector, first soak it for several hours in Gunk or diesel

fuel to loosen everything up. Bosch stipulates a *minimum* of four hours. Two ounces of caustic soda dissolved in one pint of water with half an ounce of detergent will go to work on carbon deposits. This volume of caustic soda must not be exceeded or corrosion is possible. The parts to be decarbonized are boiled in this solution for an hour, continually topping up the water to compensate for any that evaporates. Before you reassemble the injector, it will need a thor-

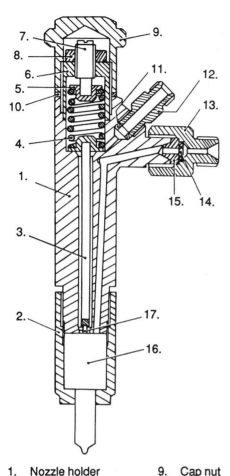

1. Nozzle holder
2. Nozzle nut
3. Spindle
4. Spring
5. Upper spring plate
6. Spring cap nut
7. Spring adjusting screw
8. Locknut
9. Cap nut
10. Joint washer
11. Joint washer
12. Leak-off adaptor
13. Inlet adaptor
14. Filter
15. Nipple
16. Nozzle
17. Needle

Figure 6-25. *Injector with spring adjusting screw. (Courtesy Lucas CAV Ltd.)*

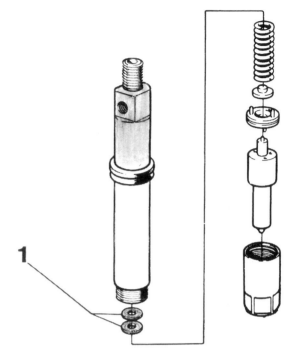

Figure 6-26. *Injector with internal shims (1) to set the opening pressure of the spring. (Courtesy Volvo Penta)*

ough flushing and drying to remove all traces of the caustic soda.

In order to disassemble an injector, hold the injector body firmly. A special vise is recommended, although it will probably not be available. The injector must not be directly clamped in a steel vise because this may distort it. Remember how precise everything is. Protective wooden blocks should be placed around the injector body and the vise given only the minimum necessary pressure. (*Note: the individual injectors on Detriot Diesels also incorporate the fuel injection pump. They are more complex than those illustrated and no attempt should be made to disassemble them.*)

Within the injector is a powerful spring (see Figure 6-25). Sometimes the spring pressure is externally adjustable (as illustrated) by removing a cap nut on the top of the injector. If this is the case, the locknut and screw should be backed off *an exact number of turns carefully counted* until the spring is no longer under tension. This spring determines the pressure at which the injector opens and is set at the factory on a special testing device. In the field without the proper equipment, the spring pressure cannot be accurately reset, so the best you can do is to put it back where it was.

On other injectors, the opening-spring pressure is set by fitting a number of shims (spacers) under the spring (see Figure 6-26). The more shims, the higher the opening pressure. Every 0.001″ (one thousandth of an inch) increase in shim thickness raises the opening pressure by about 55 psi. These injectors have no external spring adjustment, and you can move directly to removal of the nozzle assembly.

Unscrew the injector nozzle nut. Some injectors are designed to project a spray in a specific direction. In this case, the nozzle and injector body are held in the correct relationship with a small dowel. If the nozzle nut is particularly hard to break loose, the whole assembly should be soaked again because excessive force is likely to damage the dowel pin. Sometimes a sharp tap on the end of the wrench is necessary to break the grip of carbon in the injector and get the nut moving.

Once the nut is off, you can remove the nozzle and its components. The order of all the parts must be carefully noted. On no account must injectors be mixed up—nozzles and needle valves are machined as matching sets and must always go together.

Cleaning. All contact surfaces within the injector should be clean and bright. The caustic soda solution, appropriate scrapers, or both, are used to clean up carbon inside the injector. Do not scratch the needle valve, its seat, or the nozzle bore. No abrasive cleaning or grinding compounds should ever be used on the needle or its seat in the nozzle. These are machined at different angles to ensure a tight line contact (see Figure 6-27A and B). Any grinding or lapping will destroy this fit.

Injector manufacturers sell nozzle-cleaning kits (see Figure 6-28). The following are the instructions issued by CAV on the use of their kit (CAV workshop manual #C/P 24E "Fuel Injectors," page 7):

"To clean the nozzle of a multi-hole injector with the tool kit:

1. Remove all traces of carbon deposit from the exterior of the nozzle [refer to Figure 6-29] and from the needle with the wire brush. Polish the needle with a piece of soft wood; do not use an abrasive cleaning compound.
2. Clean the gallery with the scraper.
3. Clean the nozzle seat with the scraper, using the appropriate end of the scraper.
4. Similarly, clean the cavity with the scraper.
5. Use the pin vise with the appropriate size of pricker wire to clean the spray holes in the nozzle tip. The vise must be used carefully to avoid the risk of breaking the pricker wire in a spray hole."

A clean needle valve should drop easily into its seat and fall back out when the injector is inverted. Injector parts must be spotlessly clean (*lint-free* rags or paper towels only) and dipped in clean diesel before reassembly.

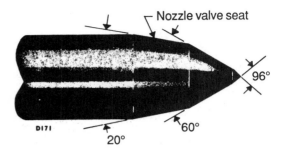

Figure 6-27A. *Nozzle valve seat. (Courtesy United Technologies Diesel Systems)*

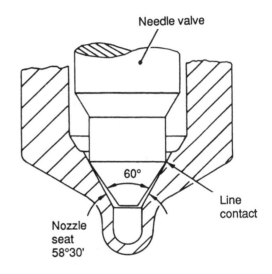

Figure 6-27B. *Angular difference between nozzle seat and needle valve (multi-hole injector). (Courtesy Lucas CAV Ltd.)*

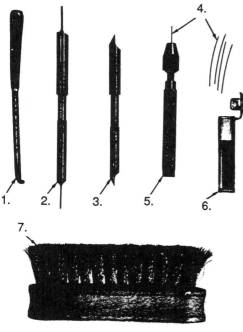

Nozzle cleaning kit

1. Scraper (gallery)
2. Scraper (cavity)
3. Scraper (nozzle seat)
4. Pricker wires
5. Pin vice for pricker wires
6. Container (pricker wires)
7. Brass wire brush

Figure 6-28. *Nozzle-cleaning kit. (Courtesy Lucas CAV Ltd.)*

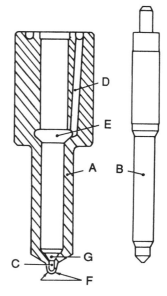

Cleaning points

A. Nozzle   E. Gallery
B. Needle   F. Spray holes
C. Cavity   G. Nozzle seat
D. Feed hole

Figure 6-29. *Cleaning the nozzle and needle. (Courtesy Lucas CAV Ltd.)*

Reassembly. This process is the reverse of the disassembly procedure. The things that need to be checked, such as the nozzle lift and spring pressure, are beyond the scope of an amateur. This is why, if at all possible, injectors should be left alone. The opening pressure adjusting screw (if fitted) should be screwed down to its previous position, as already noted. All shims must be put back exactly as before, in the same order, with a thick spacer on the top and the bottom of the thin spacers. Now, you can check the spray pattern again.

If you have lost the spring setting of an adjustable spring, turn the adjusting screw until it is finger-tight.

Thereafter, every turn represents 900 to 1,000 psi opening spring pressure. Multi-hole injectors are generally set to around 2,200 psi (2 to $2^{1}/_{2}$ more turns), and pintle nozzles to around 1,500 psi ($1^{1}/_{2}$ more turns). If the spray pattern is still poor, it may be improved by tightening the spring pressure by up to $^{3}/_{4}$ of a turn more, but certainly not beyond this. If after all this the needle valve fails to seat properly and the injector refuses to operate correctly, you can do nothing short of replacing the nozzle and needle valve assembly, or preferably, the whole injector.

An injector must make a gas-tight seal in its cylinder head. The hole in the head must be clean; a new sealing washer (if one is used) should be fitted if at all possible (if not, see "Annealing" below). If the injector is the type that's held down with a steel plate, the plate must be squarely seated and the hold-down bolts torqued evenly. A smear of high-temperature grease (e.g., "NeverSeize") around the injector barrel will prevent corrosion from locking the injector in the cylinder head.

Fuel lines are specifically made for individual cylinders, and must be returned to these cylinders. The lines will make an exact fit: both ends should be put in place at the same time then both hand-tightened before final tightening. You should never have to bend the line or force it to fit. Be especially careful with internal fuel lines since any bending may cause the line to rupture at a later date, allowing diesel fuel to dilute the engine oil. If a fuel line needs replacing, you must buy the correct individual line from the engine manufacturer. Fuel line nuts must not be over-tightened, because this distorts and fractures flare fittings—tighten most nuts to 12 to 15 lb. ft. (16-20 Nm).

Always check a fuel system for leaks after you have removed and replaced an injector or fuel pipe. Where there are internal fuel lines, run the engine for 20 to 30 minutes then remove the valve cover. Check the cylinder head and the various lubricating oil puddles for any signs of diesel leakage.

Annealing Washers. Injectors are often sealed in cylinder heads with soft copper washers. These may also be used on other parts of the fuel system. After being subjected to high temperatures, or simply over time, these washers become hard, and if you reuse them, they lose their sealing properties. Hardened copper can be readily softened by heating it to a cherry red color with a propane torch (or over a propane stove or primus), then dropping it into cold water. This is known as *annealing*. (Copper-based metals are annealed by rapid cooling, whereas iron-based metals are annealed by slow cooling—rapid cooling of iron induces hardness.)

Electrical Equipment

Dead batteries are the most common electrical failure in marine use. As often as not this comes from a lack of understanding of the unusual electrical operating conditions found on most boats, which require the use of deep-cycle batteries and special charging equipment. Chapter 4 described how to troubleshoot an engine's starting circuit; Chapter 9 goes into more detail on how to set up the electrical system in the first place so that it doesn't cause problems. The following

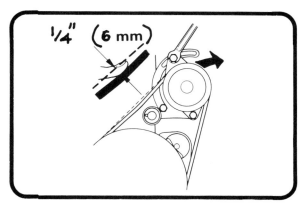

Figure 6-30. *Checking V-belt tension. (Courtesy Volvo Penta)*

are a few comments on likely causes of alternator failures.

If the charging circuit is not operating, first check the alternator drive belt—it may be broken or slipping. You should not be able to depress the belt by more than $1/4''$ to $1/2''$ with moderate finger pressure in the middle of the longest belt run (see Figure 6-30). A slipping belt often produces a rhythmical squealing, especially immediately after start-up; the output (if you have an ammeter) will cycle up and down. Tighten the belt immediately. If it is left to slip, it will heat up the belt and the alternator pulley. The former will make the belt brittle and prone to rapid failure. The latter can sometimes cause the alternator shaft and rotor to become hot enough to demagnetize the rotor, which will almost completely disable the alternator's output for good.

If the alternator circuit to the battery is broken at any time while the alternator is running, it will almost instantly cause the diodes in the alternator to blow out. Most boats are fitted with a battery selector/isolation switch. If this switch is opened (the batteries isolated) while the engine is running this will frequently blow out the diodes. Even switching from one battery to another is likely to blow the diodes, unless the switch is of the *make-before-break* variety. This type of switch first brings both batteries on line then disconnects one of them so that the connection to the boat's circuits is never interrupted.

Connecting battery leads in reverse at any time, or

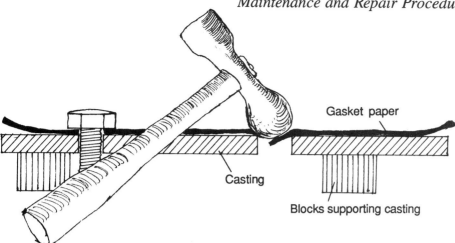

Figure 6-31. *Making a gasket. Using the head of a carriage bolt works well (bottom).*

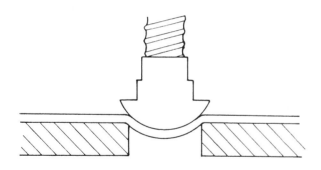

accidentally grounding the alternator's output wire when it is running, will also probably destroy the diodes. Using a fast charger on the battery with the battery still connected to the boat's circuits does the same.

High-output alternators are becoming increasingly common in boats (see Chapter 9). The higher the output, the more massive the cables have to be to carry the load. Heavy cables mounted to the back of an alternator tend to vibrate loose. This generates tremendous arcing, which can rapidly melt down the output stud on the alternator, at which point the cable drops off. As long as the battery isolation switch is not tripped, the cable is likely to have a direct connection to the battery. It will create a dead short if it touches any grounded surface (e.g., the engine block) where it

may then start a fire. *Be certain to lock the cables to the alternator, preferably with locknuts, and to check this connection for tightness at periodic intervals.*

For more comprehensive troubleshooting and repair procedures on alternators, see my *Boatowner's Mechanical and Electrical Manual.*

Gaskets

Sometimes you may have to improvise a gasket in the field. Your repair kit should contain a roll of high-temperature gasket material, plus some cork or rubber-based material for valve cover and pan (sump) gaskets. Most other gaskets can be made from brown paper, if necessary using several layers.

Even complicated gaskets are relatively simple to make. The trick is to lay a sheet of the gasket material over the piece that needs the gasket. Using the end of a ball-peen hammer (see Figure 6-31), lightly tap the gasket into all the bolt holes. The relatively sharp edges of most castings will cut the gasket material, making a perfect hole. Now slip a bolt through the gasket to hold it in place. Repeat this procedure at a couple more widely spaced bolt holes to hold the sheet of gasket paper securely in place. Tap out the remaining holes or other areas until you complete the gasket. The keys to success are striking the gasket in a way that forces it against the sharpest edge of the piece being gasketed, and using the minimum amount of force necessary (especially on aluminum) to avoid burred edges, cracked castings, and other damage. A flurry of light taps is far better than a heavy blow.

If you don't have a ball-peen hammer, you can use a box-end wrench, an upside-down carriage head bolt, or any other curved metal object. Where the outline of the piece being gasketed is not sharp enough to cut the gasket paper, rub an oily finger over it. This leaves a clear enough line to follow with a knife or a pair of scissors.

Winterizing

Certain aspects of laying up a boat for the winter have already been covered. Here is a quick recapitulation and a few additional points.

1. All raw-water systems must be drained, with particular attention to low spots. Remove the impeller from flexible-impeller raw-water pumps then run the engine for a second or two to drive remaining water out of the exhaust system. Lightly lubricate the pump impeller with dish-washing liquid and replace it, leaving the cover loose. (Put a conspicuous notice somewhere to remind yourself that you have done this.) Alternatively, close the sea-water intake seacock, disconnect the raw-water pump suction line, place it in a bucket of 50% water mixed with 50% antifreeze, and run the engine until the solution is pumped out of the exhaust.

2. Break loose the exhaust and check for carbon accumulation. Check the raw-water-injection elbow for corrosion.

3. Check the antifreeze in closed cooling systems and preferably renew it.

4. Remove the battery from the boat and put it on a *float* charge.

5. Check the sediment bowl of a primary fuel filter for any water and take a sample from the bottom of the tank, which should be pumped out as necessary. Filling the tank will reduce the volume of air in it, cutting down on condensation.

6. *Change the engine oil at the beginning of winter, not the end.* Diesel engine oils build up corrosive acids and some water over time. You don't want these going to work on the bearings all winter long.

7. Squirt a small amount of oil into the air-inlet manifold if at all possible and turn over the engine (without starting) to draw the oil into the cylinders and spread it around the upper cylinder walls.

8. Grease all grease points.

9. Remove the inner wires of all engine controls from their outer sheaths, clean, inspect, grease, and replace (see Chapter 8).

10. Inspect all flexible feet and couplings for signs of softening (from oil and diesel leaks).

11. Inspect all hoses for signs of softening, cracking, bulging, or all of these, especially hoses on the hot side of the cooling system and in the exhaust. Rinse out anti-siphon valves in fresh water. Undo all hose clamps a turn or two to check for crevice corrosion under the clamp, then retighten.

12. Seal all openings into the engine (air inlet, breathers, exhaust) and the fuel tank vent. *Put a conspicuous notice somewhere that you have done this so that you remember to unseal everything at the start of the new season or serious damage may result.*

When recommissioning the engine in the spring, you will need to refill the raw-water system (if it was drained), bleeding the pump and piping as necessary; tighten down the cover on the raw-water pump; replace the battery; and unseal all the openings into the engine. Then you must crank the engine for 30 seconds, with the throttle closed, to relubricate all the bearings before you start it. This is particualrly essential on supercharged engines.

Chapter 7

Maintenance and Repair Procedures, Part Two: Decarbonizing

Sooner or later carbon build-up in the cylinders, and in the cylinder head, will necessitate decarbonizing and doing a valve job. Decarbonizing consists of removing the cylinder head, and perhaps the pistons, and removing baked-on carbon from them and all associated parts (valves, etc.).

In the case of engines whose valves are operated by push rods, this is well within the capability of an amateur mechanic and should give no cause for alarm. As in all other areas of maintenance, as long as you keep the work area and engine clean and approach the job calmly and methodically, you shouldn't encounter any insurmountable problems.

On engines with an overhead camshaft (a camshaft in the cylinder head eliminates the need for push rods), removing the cylinder head to decarbonize will disturb the valve (and possibly the injection pump) timing. At the end of this chapter the general principles that govern pump and valve timing are described, but a manufacturer's manual may be necessary to learn the specific procedure for timing a particular engine.

When should you decarbonize? There are two schools of thought: 1) perform all maintenance at preset intervals, or 2) wait until you have problems. The logic of the former position is that preset maintenance intervals will deal with conditions likely to cause problems before they get out of hand. The logic of the latter position is that no two engines operate under the same conditions and one engine may run five times longer than another before it needs decarbonizing, or any other substantial overhaul. I lean toward the latter.

"If it ain't broke, don't fix it" is not such a bad idea, as long as you are religious about routine maintenance, especially oil and filter changes, and ensuring clean fuel. This alone will go a long way toward lengthening the intervals between substantial overhauls. The next most important thing is to avoid prolonged running at low loads.

Once a problem becomes evident (smoky exhaust, loss of power, difficult starting, etc.), you must rapidly deal with it. If these conditions are caused by valve or piston blow-by, for example, the hot gases won't take long to do some serious damage.

115

Preparatory Steps

The following must be done before the cylinder head can be removed:

1. The engine needs to be clean. Any time an engine is opened up, all kinds of damaging dirt will fall into it if you haven't cleaned it.

2. As a general rule, everything that comes off an engine should go back in the same place and in the same way. This is especially important for pistons, valves, push rods, rockers, and other moving parts. All these parts wear where they rub on mating surfaces. If they are switched around on reassembly, a high spot on one part may now rub against a high spot on another part, and overall engine wear will greatly accelerate.

   Set aside a clear space and protect it with newspapers or a clean cloth. As you remove pieces from the engine, spread them out in the correct relationship to one another so that you won't be confused on reassembly (see Figure 7-1). Finding a suitable space on a boat is sometimes hard, but it should be done.

3. Drain the engine of coolant to at least a level below the cylinder head. A drain valve or plug should be located somewhere at the base of the block (see Figure 7-2). Since most sailboat engines are below the waterline, the engine-water seacock on a raw-water-cooled engine must first be closed. On engines with a heat exchanger and a header tank, the radiator cap on the header tank needs to be loosened to break the vacuum that forms when you drain the block.

4. All equipment attached to the cylinder head has to be removed. This includes inlet and exhaust manifolds, a turbocharger, and after-cooler if fitted (see Figure 7-3).

5. All injection lines (delivery pipes) have to be

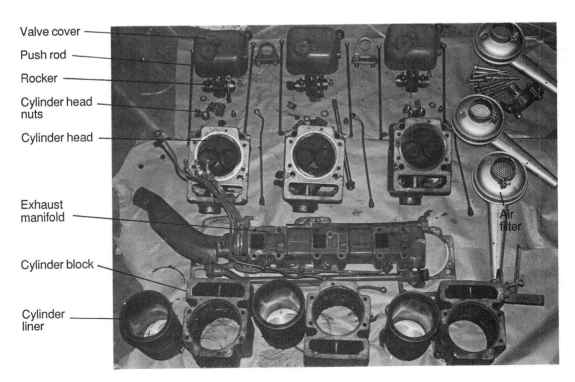

Valve cover
Push rod
Rocker
Cylinder head nuts
Cylinder head
Exhaust manifold
Cylinder block
Cylinder liner
Air filter

Figure 7-1. *Keep things clean, tidy, and in the correct relationship to one another.*

Figure 7-2. *Cylinder-block drains in a Volvo MD17C.*

broken loose from their respective injectors (or the fitting on the outside of the cylinder head if they are internal), and from the injection pump (see Figure 7-4). *The minute any fuel line is disconnected, you must cap it and the unit to which it is attached to prevent the entry of ANY dirt into the fuel system* (see Figure 7-5). Number the fuel lines for ease of reassembly—a piece of masking tape and a felt-tip pen work fine. If the injectors are to be overhauled, taking them out now is easier than when the cylinder head is off (see the section on injectors in Chapter 6).

6. Remove the valve cover and unbolt the rocker assembly (see Figures 7-6 and 7-7). On some engines, it comes off in one piece; on others, each cylinder has a separate unit (see Figure 7-9). The push rods can now be taken out and laid down in order (perhaps labeled, as with the injector lines).

7. On some engines with an overhead camshaft, gears drive the camshaft; on others it is powered by a belt or chain (see Figure 7-8). To remove the rockers or the cylinder head, you must first remove the belt or chain. Before you can do that, you must remove the crankshaft pulley and the timing case cover on the front of

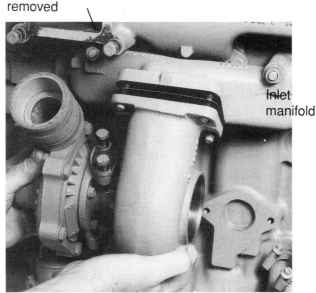

Exhaust manifold—an intermediary section of exhaust ducting has already been removed

Inlet manifold

Figure 7-3. *Removing a turbocharger from the inlet and exhaust manifolds. (Courtesy Perkins Engines Ltd.)*

the engine. You will probably need a special puller to remove the pulley.

If the pulley has any tapped (threaded) holes in its face, you can make a puller (see

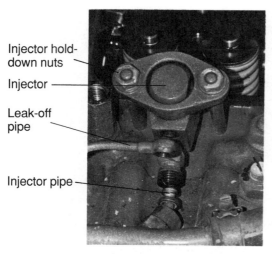

Injector hold-down nuts

Injector

Leak-off pipe

Injector pipe

Figure 7-4. *Cylinder head with injector lines broken loose.*

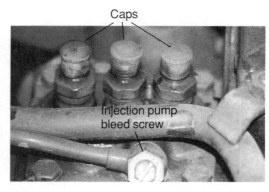

Caps

Injection pump bleed screw

Figure 7-5. *Fuel lines properly capped. (Volvo MD17C injection pump)*

Figure 7-10). Slack off the pulley's retaining nut. It may have a lock washer with a tab that must first be bent back out of the way. The nut may be difficult to break loose because the engine will turn over when you apply pressure to the wrench. If a smart blow on the wrench fails to jar loose the nut, grasp both sides of the water pump, or alternator, drive belt, hold it tightly, and try again. If the boat has a manual transmission, place the engine in gear and lock it by putting a pipe wrench on the propeller shaft—wrap the shaft with a rag first in order to avoid scoring the shaft.

Now, bolt a flat metal plate, with either a tapped hole drilled in its center, or a nut welded over the hole, across the pulley, using the threaded holes in the face of the pulley. Screw a bolt down through the tapped hole or welded nut onto the end of the crankshaft. This will drive the pulley off the shaft (see Figure 7-10). (Note: if the nut is placed on the other side of the plate, as long as you can get a wrench in to hold it, it doesn't need to be welded in place).

Any time you remove a camshaft, the fuel injection pump timing and the valve timing will be disturbed and will need resetting. This

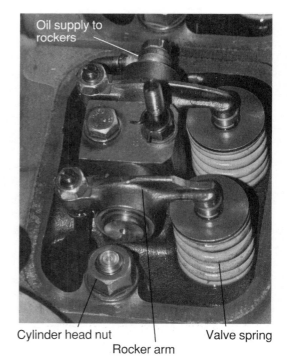

Oil supply to rockers

Cylinder head nut

Rocker arm

Valve spring

Figure 7-6. *Engine with valve cover removed.*

timing is absolutely critical to engine operation. If you have any doubt about being able to reset either valve or injection pump timing, do not disturb them. You are now ready to remove the cylinder head.

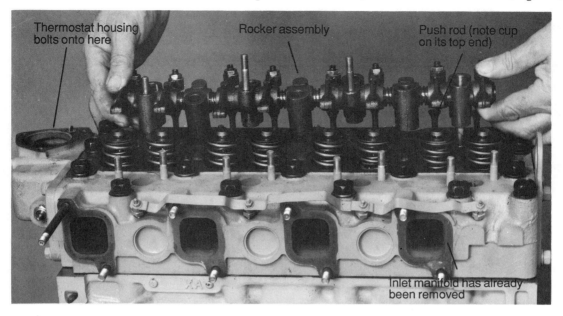

Figure 7-7. *Lifting off the rocker assembly. (Courtesy Perkins Engines Ltd.)*

Figure 7-8. *Overhead camshaft.*

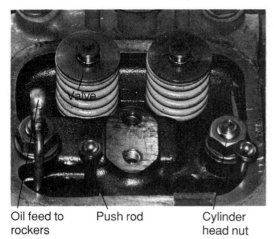

Oil feed to   Push rod   Cylinder
rockers   head nut

Figure 7-9. *Cylinder head with rockers removed.*

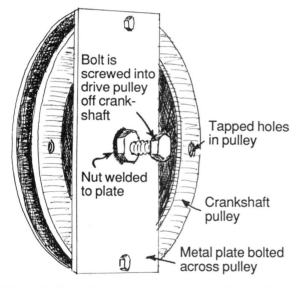

Figure 7-10. *A homemade crankshaft pulley puller.*

Cylinder Head Removal

A cylinder head is held down by numerous nuts or bolts spaced around each cylinder. In order to evenly relieve the pressure exerted on the cylinder head, loosen each nut (or bolt) a half turn or so in the sequence outlined in the engine manual. If no manual is available, you should loosen a nut (or bolt) at one end of the head a half turn, then loosen the one opposite it at the other end, and then another at the first end, and so on, working toward the center of the head. After you have released the initial tension, you can remove all the nuts (or bolts).

A cylinder head will frequently bond to the cylinder block, making it difficult to break loose. When this happens, the temptation to stick a screwdriver in the joint between the two and beat on it with a hammer is dangerous and must be resisted. The tremendous pressures concentrated on the tip of the screwdriver can result in a cracked head or block. Instead, if the injectors are still in place try turning the engine over; the compression will often be enough to loosen the cylinder head. Failing this, firmly hold a solid block of wood to the head at various points and give it a moderate whack with a hammer or mallet. The shock should be enough to jar the head loose. Make sure that the wood contacts a good area of the head; any point loading can crack the head. Be sure to give the wood a smart blow. If the head still refuses to budge, check to see that all the fastenings have been removed—it is surprisingly easy to miss one, especially on dirty engines.

When the head starts to come up, you must lift it squarely in order to avoid bending the hold-down studs (see Figure 7-11). Do not drag it across the top of the studs because you may scratch the face of the head. Sometimes you can remove the head with the rocker assembly still in place and the push rods still in the engine. If you do this, the push rods may stick to the rockers when you lift off the head. On a few engines they may then drop into the oil sump, and you'll have to remove it to retrieve them! If any push rods stick, lift the head an inch or so then carefully replace them on their cam followers before proceeding.

The injector tips of open-combustion (direct injection) engines often protrude below the level of the cylinder head (see Figure 7-12). Do not rest the head on them, as this will damage the nozzles.

Any time you remove a head, or any other piece of equipment, remember to block off all exposed passages and holes into the engine to prevent trash and engine parts from falling inside. It is unbelievably frustrating to drop a nut down an oil drain passage and into the sump then have to remove the engine from the boat in order to drop the sump and recover the nut.

Once the head is off (see Figure 7-13), remove old gasket material and other dirt from the face of the head and the block. Gaskets frequently become extremely well bonded to cylinder heads and blocks.

Figure 7-11. *Lifting off a cylinder head. (Courtesy Perkins Engines Ltd.)*

Figure 7-12. *A cylinder head removed (Volvo MD17C). Do not rest the head on the injector nozzle. (Note: This engine has a direct combustion chamber; therefore, it doesn't have a precombustion chamber.)*

Figure 7-13. *A Volvo MD17C with cylinder heads removed. Note the toroidal crown of the pistons, which is common to direct-combustion engines.*

You can buy a variety of scrapers from automotive stores, or you can use a paint scraper. An old chisel (about 1″ wide) or any good-size pocket knife will do. The key, especially on aluminum, is to keep the blade at a shallow angle to the surface being cleaned, otherwise you risk scratching or gouging the metal. Particularly stubborn residues require patience. On cast iron they can be softened with paint remover, but chemicals are not advised on aluminum.

With the head off and its face cleaned, now is probably a good time to check the head for warpage. Lay a straightedge (a steel ruler is excellent) across it at numerous points and attempt to slide a *feeler gauge* under it (see Figure 7-14). Feeler gauges are available from any automotive parts stores. They consist of a number of thin metal blades, precision ground to the specified thickness stamped on the blade face. A set from 0.001″ to 0.025″—one thousandths of an inch to twenty-five thousandths, or the metric equivalent if you have a metric engine—is needed. See Appendix C for conversion tables.

Allowable warpage varies according to the size of the cylinder head. If the manufacturer's specifications are not available (as they almost certainly will not be), you can safely assume that on the engine sizes under consideration here, any warpage over 0.004″ to 0.005″ (four to five thousandths of an inch) is excessive.

Valves

You can test valves for leakage by laying the head on its side and pouring kerosene or diesel into the valve ports. If a valve is bad, the liquid will dribble out where the valve rests on its seat. If no leak is present, you may be wise to leave the valve in place and merely clean the carbon off its face and out of its port.

Valve Removal. If leakage is present, the valve will have to be reground. Valves are usually held in place by two *keepers* (or *collets*), small semi-circles of metal that lock in a slot cut into the valve stem (see Figure 7-15). A dished metal washer on top of the valve spring holds the keepers against the valve stem. In order to remove the keepers, you must compress the valve spring far enough to allow you to push the dished washer down out of the way. A valve spring compressor or clamp is used to do this. It is essentially a large C-clamp that fits over the cylinder head, one end resting on the face of the valve and the other slotting over the valve stem and around the top of the spring (see Figure 7-16). Closing the clamp compresses the dished washer and spring down the valve stem, allowing the keepers to be picked off. Releasing the clamp permits the washer and spring to slide off the valve stem and the valve to be pushed out of the cylinder head.

Some automotive parts stores rent valve-spring clamps. Most are sufficiently adjustable to fit a wide range of cylinder heads, but if at all possible, take the head to the store and check the clamps for the best fit.

In an emergency, it is possible (though not easy) to push down on the valve spring with a suitably-sized box-end wrench, allowing a second person to remove

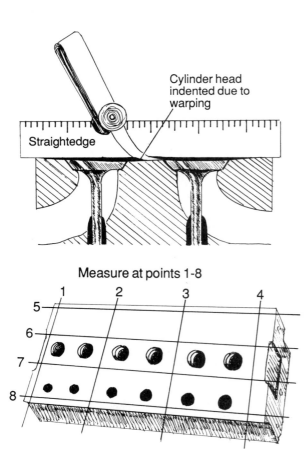

Measure at points 1-8

Figure 7-14. *Checking for cylinder-head warpage.*

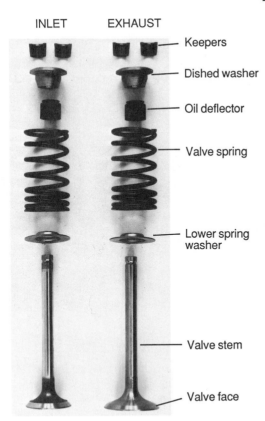

INLET   EXHAUST
— Keepers
— Dished washer
— Oil deflector
— Valve spring
— Lower spring washer
— Valve stem
— Valve face

Figure 7-15. *Inlet and exhaust valve components. (Courtesy Perkins Engines Ltd.)*

Figure 7-16. *Top: Using a valve-spring compressor. (This is slightly different from the one described in the text, but it serves the same purpose.) (Courtesy Perkins Engines Ltd.) Bottom: A hydraulically operated valve-spring compressor in a large machine shop. (Courtesy Caterpillar Tractor Co.)*

the keepers. This is easier with older, slower-revving engines, which tend to have weaker valve springs. The trick is to pick off the keepers without allowing the wrench to slip off the spring, letting it shoot off the stem. The tiny keepers often get lost when this happens. Keepers are hard to buy and easy to lose—handle them with care.

Cleaning and Inspection. A cast-iron cylinder head (but *not* aluminum) and all the steel parts (valves, springs, etc.) can now be degreased in a solution of one pound of caustic soda plus eight ounces of detergent dissolved in a gallon of water. If possible, put the solution in a tub large enough to hold the cylinder head, and heat the whole thing for an hour or so. Afterward, be sure to flush the head very thoroughly with copious amounts of fresh water.

The key area of a valve is the beveled region that sits on the valve seat in the cylinder head (see Figure 7-15). Inspect valves for any of the problems illustrated in Figures 7-17, 7-18, and 7-19. If the valve or seat is pitted in the area of contact, the head will have to go to a machine shop for regrinding of the seat and refacing of the valve (see Figure 7-18). Exhaust valves need checking more closely than inlet valves because they are subject to much higher temperatures, and the exhaust gases tend to burn them more quickly. The exhaust valve closest to the exhaust manifold exit pipe is frequently the most corroded on a marine engine as a result of water vapor from a water-cooled exhaust coming back up the exhaust pipe.

If the seat and valve face are reasonably smooth, you can frequently *lap* in the valve by hand, as discussed in the next paragraph, although doing this to the specially hardened valves and seats increasingly common in modern engines is less feasible than it used to be. Lapping hardened valves and seats by hand may alleviate minor problems, but problems that cannot be solved in this fashion will likely require new valves and probably new seats since machining either in any way will cut through the hardened surfaces.

Lapping. To lap in a valve, apply a thin band of medium grinding paste (available from any automotive parts store) around the seating surface. Drop the valve into the cylinder head, and place a lapping tool (also available from any automotive parts store) on the

Figure 7-17. *Bottom left: Some cupping of valve faces is normal, but if it's excessive (more than .010"), as it is here, the valve must be discarded. Bottom: Use a straightedge and feeler gauge to check for valve cupping. Top right: If the valve has large or deep marks near the edge, discard it. (Courtesy Caterpillar Tractor Company)*

Figure 7-18. *Valve failure. Top left and right: corrosion from moisture and acids. Bottom right: stress cracks caused by high temperature. Bottom left: metal-to-metal transfer (galling) from a valve stem's sticking in its guide. (Courtesy Caterpillar Tractor Co.)*

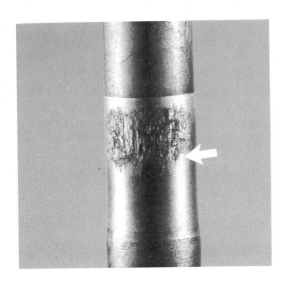

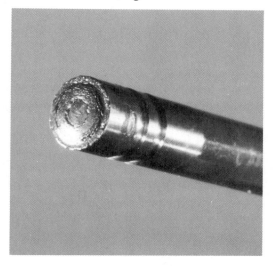

Figure 7-19. *Valve failure. Top left: valve badly burned by escaping gases. Top right: beaten and battered valve stem end from excessive tappet clearance. Bottom left: damage caused by a foreign object bouncing around in the combustion chamber. Bottom right: a slightly bent valve. Notice the uneven grind marks on the face. (Courtesy Caterpillar Tractor Co.)*

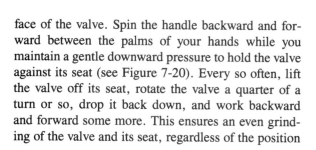

face of the valve. Spin the handle backward and forward between the palms of your hands while you maintain a gentle downward pressure to hold the valve against its seat (see Figure 7-20). Every so often, lift the valve off its seat, rotate the valve a quarter of a turn or so, drop it back down, and work backward and forward some more. This ensures an even grinding of the valve and its seat, regardless of the position

of the valve. (Some valves have a screwdriver slot in them, in which case the lapping tool is unnecessary.)

Continue this procedure until a line of clean metal

is visible all the way around the valve and its seat. Add more grinding paste as you need it. As soon as this line appears, polish the surfaces in the same manner, using a little fine grinding paste.

You can check the fit of a valve in its seat by making a series of pencil marks across the face of the valve about $1/8''$ apart then dropping the valve onto its seat. All the pencil marks should be cut by the seat (see Figure 7-21).

Do not overgrind the valves. The objective is a thin line of continuous contact between the valve and its seat—not a perfect fit. Overgrinding lowers the valve in the head, which increases the size of the combustion chamber and leads to a loss of compression. Once the valves and seats have been ground beyond a

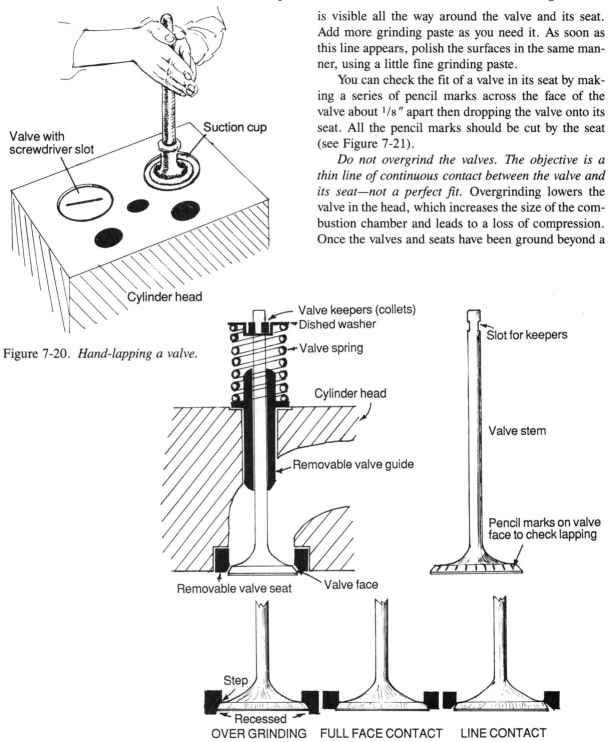

Figure 7-20. *Hand-lapping a valve.*

Figure 7-21. *Checking the fit of a valve in its seat.*

certain point, the loss of compression becomes unacceptable and necessitates a major cylinder head overhaul. The valve seats of some engines are separate inserts pressed into the cylinder head, and they can be removed and replaced. Combined with new valves and valve guides (see below), this produces, for all intents and purposes, a new cylinder head. This work can only be done by a qualified machine shop.

Valve Guides.

A valve guide holds the valve in alignment in its cylinder head. A build-up of carbon around the valve stem and in the guide sometimes causes a valve to stick in its guide. These areas must be carefully cleaned during decarbonizing. Excessive wear in a guide allows lubricating oil from the rocker arm to run down the valve stem into the valve port, where it is sucked into the engine and burned (in the case of an inlet valve), or burned by the hot exhaust gases (in the case of an exhaust valve). Check wear on the valve guides and stems by attempting to rock the valve from side to side in its guide. No lateral movement is permissible—the wear limit is normally around 0.005″ but requires a dial indicator to measure. Most engines have replaceable valve guides, which are pressed into the cylinder head, but this is a job for the machine shop.

Valve Replacement.

Before you replace any valves, visually check the springs for cracks, nicks, and corrosion (see Figure 7-22). Check the length of each spring against the manufacturer's specifications, if possible, or compare it to a new spring. Replace the spring if it is damaged or short.

Replacement of valves is the reversal of removal. First, wash away all traces of grinding paste by thoroughly flushing the cylinder head and components with diesel or kerosene. After you've refit the valves, do the kerosene test to check their seating.

B

A

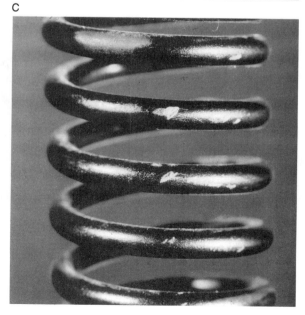

C

Figure 7-22. *(A) Wear on the sides of a valve spring is not acceptable; discard it. (B) Notches at the end of a valve spring are not acceptable; discard. (C) If a valve spring has deep nicks or notches, it must be discarded. (Courtesy Caterpillar Tractor Co.)*

Figure 7-23. *Checking for overgrinding of the valves. (Courtesy Perkins Engines Ltd.)*

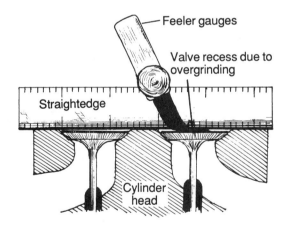

Figure 7-24. *Cutaway view of checking valves for overgrinding.*

At this point, you should probably test older engines for overgrinding of the valves. To do this, lay a straightedge across the face of the cylinder head, over the top of the valve, and measure with a feeler gauge the extent to which the valve is recessed into the head (see Figures 7-23 and 7-24). Check the degree of valve indentation against the manufacturer's specified limits to determine if the head needs new valves and seats.

Cylinders

Cylinder Inspection. When a piston is at the top of its stroke, its topmost ring is still a little way down the cylinder. Because a cylinder wears only where it is in contact with the rings, this top part of the cylinder will be unworn. A significantly worn cylinder bore will have a step right below this. In order to check for wear, rotate the crankshaft until the piston is at, or near, the bottom of its stroke. Clean away the carbon that has collected at the top of the bore with a suitable scraper or knife, finishing off with a piece of wet-and-dry sandpaper dipped in diesel. Wipe the cylinder wall clean and run your fingernail up and down the first half-inch of the bore. If the step approaches anywhere near the thickness of a fingernail, you should have the bores professionally measured to determine whether the time has come for a cylinder renewal, which also includes new pistons and rings (see page 136 for a simple technique to approximately measure cylinder wear).

Other indications that the engine needs a new cylinder are any cracks, however small, or evidence of holes in the cylinder wall (such as erosion on the top flange of a wet cylinder liner). Air filter failure will cause vertical scratches in the cylinder wall. Improper injection, resulting in fuel hitting the cylinder, will burn away the affected area. If the engine has experienced a piston seizure, the softer aluminum of the piston (most pistons are aluminum) will frequently peel off and stick to the cylinder wall. All these problems will probably also necessitate a new cylinder (see Figure 7-25).

Cylinder Honing. Long hours of low-load running often results in a cylinder wall becoming *glazed*—very smooth and shiny. If there is very little other wear, glazing can be broken up by a process called honing, using a flexible nylon brush with an abrasive material on the tips. This is known as a flex-hone. Wet liners can be pulled out of the block and taken to a machine shop; dry liners can be honed with the block and cylinder still in the boat, but first the pistons must be removed (see below). The hone will generate a good bit of abrasive dust, so cover the crankshaft below the cylinder with a large rag.

You will probably have to call in a machine shop, but you may be able to rent an appropriately sized

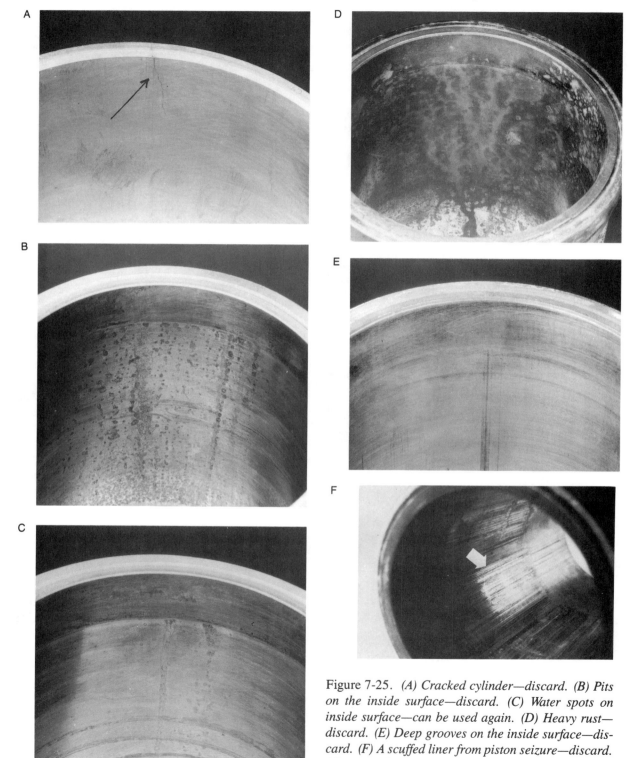

Figure 7-25. *(A) Cracked cylinder—discard. (B) Pits on the inside surface—discard. (C) Water spots on inside surface—can be used again. (D) Heavy rust—discard. (E) Deep grooves on the inside surface—discard. (F) A scuffed liner from piston seizure—discard. (Courtesy Caterpillar Tractor Co.)*

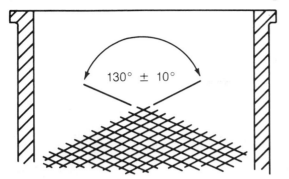

Figure 7-26. *Correct cross-hatch pattern produced by honing—no pits, rust, scratches, or shiny areas. (Courtesy Caterpillar Tractor Co.)*

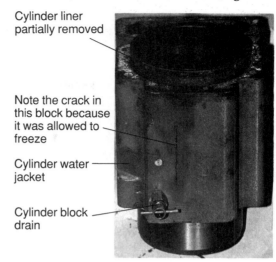

Figure 7-27. *Wet cylinder liner.*

hone. These used to be available through automotive parts stores but are pretty hard to find now. Put the hone into a slow-speed electric drill (350 to 500 rpm), lubricate it with kerosene, and run it up and down inside the cylinder at the rate of approximately 30 strokes a minute (one second down; one second up). *On no account hold the hone stationary*—it will score the cylinder wall. The idea is to produce over the entire piston a cross-hatch pattern of very light scratches angled at around 130 degrees to one another (see Figure 7-26)—and covering the entire piston-ring contact area of the piston. The angle of the cross-hatch can be adjusted by changing the speed of the drill or the rate at which you move the hone up and down the cylinder. If the slope of the cross-hatch is less than 130 degrees, increase the speed of the drill or decrease the rate at which you move it up and down. If the angle is more than 130 degrees, slow the drill or increase the rate of movement. Use the flex-hone for approximately 30 seconds only, certainly no more than 60 seconds —the time will depend on how badly glazed the cylinder is and how worn-out the hone is.

After you hone the cylinder, thoroughly clean it. Swab it down with kerosene and wipe it with paper towels until the towels come up spotless. Immediately after cleaning, wipe the cylinder with a towel soaked in oil—rust can begin to form in minutes on a clean, dry cylinder wall. After you remove the rag from the crankcase, thoroughly flush the area to remove all traces of the abrasive material shed by the hone.

If a wet liner (see Figure 7-27) has been removed

from the engine, fit new O-ring seals before replacing it. Lubricate the seals with liquid soap (dishwashing liquid works well). If the liner has a light scale or corrosion on just one side, rotate it 90 degrees from its previous position when you re-install it.

Piston Removal

Pistons are sealed in their cylinders by piston rings. Areas in which significant wear occurs are the cylinder wall (see above), the outer surface of the piston

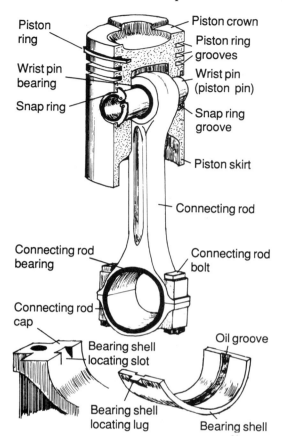

Figure 7-28. *Piston and connecting rod.*

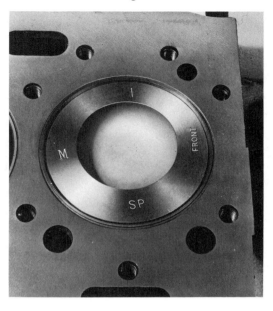

Figure 7-29. *The markings on a piston crown. (Courtesy Perkins Engines Ltd.)*

ring, and the width of the piston-ring groove. This groove widens over time as a result of the ring's working up and down as the piston moves in the cylinder. Piston rings can only be checked after a piston has been removed from its cylinder.

To remove a piston from its cylinder, you must detach its connecting rod from the crankshaft then push the piston/connecting rod assembly out through the top of the cylinder. A cap, fastened with two bolts, holds connecting rods to the crankshaft (see Figure 7-28). The majority of engines require you to remove the engine's oil pan to gain access to these bolts. In most boats, this requires lifting the engine off its bed to gain access to the pan. A few marine diesels, however, provide access to the connecting rod caps through hatches in the side of the crankcase, allowing you to remove the pistons without disturbing the engine.

Lifting an engine from its bed is a major undertaking because it involves breaking loose fuel lines, electrical connections, the exhaust system, the propeller coupling, and other equipment. You'll also need some form of overhead crane or hoist. The main boom of a sailboat, if it's long enough and adequately supported, can often be used with an appropriate block and tackle. The boom of a trawler yacht's riding sail is another possibility.

Before you take a piston from its cylinder, you have to remove the ridge of carbon at the top of the cylinder. Also, before undoing the connecting rod cap, get hold of the connecting rod where it clamps around the crankshaft journal and work the piston up and down and backward and forward. You will notice some lateral movement along the crankshaft journal, but otherwise this bearing should have no appreciable play. Play is usually accompanied by knocking, and the bearing will need replacing.

Sometimes separating connecting rod caps from their rods is difficult. *Never pry them apart.* Tap the cap gently with a soft hammer or a block of wood while you pull down on it—this will invariably break it loose.

Pistons *must* go back in the cylinder from which they came, and you must not scratch the cylinder liner. The piston must face in the same direction; the connecting rod cap must go on the same way and the connecting rod bolts into the same holes. On some engines, you must fit new bolts every time you remove the caps. This is a good practice for any engine. The piston crown (top) should already be marked with its cylinder number and forward face (see Figure 7-29), and the connecting rod and cap should also be numbered and marked. If not, you have to make some kind of identification.

Cleaning and Inspection. Pistons are cleaned commercially by using various solvents or blasting with glass beads. Assuming these are not available, a good soaking in diesel will help to loosen carbon and other deposits, which can then be removed with fine wet-and-dry sandpaper (400 to 600 grit) constantly wetted out with diesel. Do not scratch the pistons, most of which nowadays are made of aluminum. If you have cast-iron pistons, you can soak them in the caustic soda solution previously recommended for cast iron cylinder heads.

After you've cleaned the pistons, check them for excessive wear or damage. The following problems, with their likely causes, indicate that new pistons are called for (see Figure 7-30).

1. A severely battered piston crown. This is generally caused by a broken or sticking valve, or a broken glow plug or injector tip.
2. Excessive cracking of the piston crown. The crown takes the full force of combustion, and some hairline cracking is usual on modern high-speed diesels. On engines with pre-combustion chambers, this generally is concentrated at the point on the piston crown where the combustion gases drive out of the pre-combustion chamber and hit the piston. However, extensive crazing or deep cracks mean that the piston top has overheated and the piston must be replaced. The most likely cause is faulty fuel injection.
3. Parts of a piston crown may be eaten away, also as a result of faulty injection. Injector dribble causes late combustion and detonation, which leads to burned exhaust valves and erosion of pistons, especially in the area of the crown closest to the exhaust valve. The latter is a result of the continuing combustion during the exhaust cycle. A plugged air filter or defective turbocharger also causes improper combustion and can lead to more widespread burning of the piston crown.
4. A piston may become severely worn all around its sides from the crown down to the top ring. This indicates that the above problems have resulted in generalized overheating of the piston crown, causing it to expand and rub against the cylinder wall. Some scratching of this portion of the piston is normal because it rubs against the carbon ridge at the top of the cylinder. Excessive wear requires a new piston. If the piston rings are also damaged or stuck in their grooves, blow-by of hot gases is likely to spread this wear (*scuffing*) down the sides of the piston.
5. The same scuffing on the base (*skirt*) of a piston indicates widespread overheating, most likely due to a failure of the cooling system or a lack of lubrication. Left unattended, this will probably lead to a piston seizure, causing the surface of the piston to break up and stick to the cylinder wall.

Piston Rings

Piston ring failure is generally caused by poor installation practices (see below), excessively worn piston ring grooves, hitting the ridge of carbon at the top of the cylinder, water in the cylinders, or detonation as a result of improper fuel injection or excessive use of starting fluid. In all instances, the top ring is the most vulnerable.

Broken rings need replacing. Unbroken rings need checking for wear. New and old rings need checking for fit in their grooves. The rings may be stuck in the grooves with carbon and other gummy deposits, so the first task is to clean them.

Removing Piston Rings. Piston rings are extremely brittle and easily broken. You should loosen them in their grooves by carefully cleaning off excess carbon, using plenty of penetrating fluids. After you have freed the rings, the ends must be

A

B

C

D

Figure 7-30. *(A) Generalized overheating of the crown. The top of the piston side has started to peel away and stick to the cylinder wall. Piston cannot be used again. (B) Cracking of the piston crown through overheating, in this case concentrated where the gases blow down out of the precombustion chamber. (C) Severe erosion of the piston crown due to erratic combustion caused by a plugged air filter. This piston cannot be used again. The stainless steel plug in the crown, found only in high-performance engines, helps dissipate heat from the precombustion chamber. (D) Carbon cutting around the piston top from the ring of carbon at the top of the cylinder. Although the piston is scratched, it isn't breaking up and can be used after it's been cleaned. (E) The rings on this piston are stuck in their grooves, leading to overheating, blow-by, and serious erosion of the side of the piston. It cannot be used again. (F) The skirt (the area below the piston pin) of this piston has been scuffing (rubbing without lubrication) on the cylinder wall because of the engine's overheating. The skirt has begun to break up, so the piston cannot be used again. (G) This piston has been seizing from top to bottom through serious overheating or lack of lubrication. It cannot be used. (Courtesy Caterpillar Tractor Co.)*

expanded (pried apart) to enlarge the diameter sufficiently to lift them off the piston. Piston-ring expanders are available to do this and should be used if at all possible (see Figure 7-31). If you cannot lay your hands on a piston ring expander, a few strips of thin metal slipped under the ring as you expand it out of its groove should enable you to slide it off the piston (see Figure 7-32). Old hacksaw blades, carefully ground down to remove all sharp edges, work well enough. You must slide off the ring evenly—if it gets cocked, it will probably break.

While this is a simple procedure, you must take

E

F

Figure 7-31. *Piston-ring expander. (Courtesy Detroit Diesel Corp.)*

G

great care in easing the rings out of their grooves. Expand them only enough to slide them off the piston. The two ends of a piston ring often have sharp points; do not let these scratch the piston when you take off

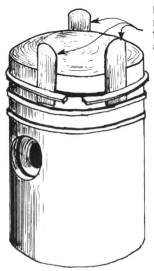

Using thin metal blades to remove a piston ring

Figure 7-32. *Piston-ring removal.*

the rings and put them on.

Incorrect removal and installation procedures are a major cause of piston ring failure. In general, rings should be removed only if strictly necessary and should then be replaced with new ones to be on the safe side.

Cleaning and Inspection.

Once you have removed the rings, clean them and the ring grooves in the piston. The latter must not be scratched or widened because this will allow gases to blow past the rings. Although it is frequently done (and I have done it many a time), I don't recommend using a piece of an old ring for cleaning out the ring grooves. Instead, make a scraper out of a piece of hardwood to fit the grooves (see Figure 7-33).

Piston rings are made of cast iron, generally with a facing of chrome where they contact the cylinder wall. Any time this chrome is worn through, the ring should be replaced. Because the action of a ring rubbing on the cylinder wall polishes its face, it is often hard to tell whether or not the chrome is gone. You can get an indication of ring wear, though, by looking at the ring's side profile. Rings are either flat-faced, tapered, rounded (*barrel-faced*), or double-faced (oil scraper rings, always the bottom ring on a piston). Tapered and barrel-faced rings contact a cylinder only at the top of the taper or barrel. As wear increases, the point of contact grows wider. If these rings are worn

flat, with the whole ring face in contact with the cylinder wall, it is time to replace them. Double-faced rings can be compared with new ones to gauge the extent of wear.

Measuring Wear.

To check piston ring wear more accurately, insert the ring into a cylinder and push it down to the bottom (no wear takes place at the bottom of the cylinder). Use an upside down piston as a plunger; it will keep the ring square, which is important for accurate measurements. Now, measure the gap between the ends of the ring (end gap), using a feeler gauge (see Figure 7-34). As a ring wears and pushes out on a cylinder wall, this gap increases. Compare the size of the gap to the manufacturer's specifications to see if a new ring is called for. As a general rule of thumb, the gap should be between 0.003″ and 0.006″ per inch of cylinder diameter.

Note that it is also possible to get some idea of the extent of cylinder wear by comparing the end gap just measured with the gap on the same ring when measured just below the lip at the top of the cylinder. The difference between the two measurements is the cylinder's increase in circumference due to wear. To convert this to its increase in diameter, the figure must be divided by 3.143.

Fit piston rings to the pistons, using a reversal of the removal procedure. Most rings have a top and a bottom—the upper face should be marked "T";

Figure 7-33. *Cleaning piston-ring grooves with a piece of hardwood. (Courtesy Caterpillar Tractor Co.)*

sometimes the bottom face is marked "BTM." In any event, when you take them off, note which side is the top and install them the same way up, and in the same grooves.

Measure the wear of the ring groove by sliding the appropriate feeler gauge into the groove between the piston and the ring (see Figure 7-35). This clearance should be about 0.003″ to 0.004″. Compare your measurements with the maker's specifications. Excessive clearance means that you will need a new piston.

Pistons are often supplied in sets, sometimes complete with connecting rods. This is to keep the engine in balance and cut down on vibration. Each piston and rod assembly is machined to the same weight as all the others in the set, and if one piston needs replacing, you may have to change all of them.

Because of the extremely close tolerances between

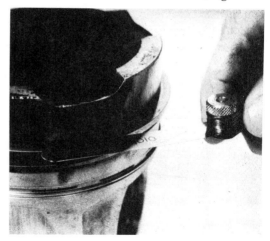

Figure 7-35. *Measuring the vertical clearance between the ring and the groove. (Courtesy Caterpillar Tractor Co.)*

Figure 7-34. *Checking piston-ring gap. Note: The ring is shown at the top of the cylinder for clarity. You should measure the ring gap with the ring at the bottom of the cylinder unless the cylinder is brand-new. (Courtesy Perkins Engines Ltd.)*

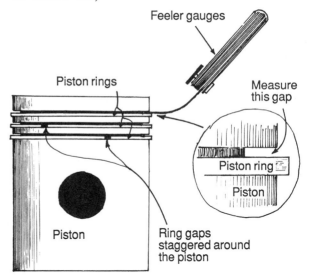

Feeler gauges

Piston rings

Measure this gap

Piston ring

Piston

Piston

Ring gaps staggered around the piston

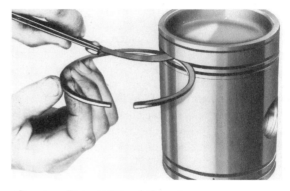

(Courtesy Detroit Diesel Corp.)

the piston crown and the cylinder head, new pistons for some older engines are made oversize and later machined in a lathe for an exact fit in the cylinder. This is known as *topping.* If you have this type of piston, this work will have to be done by a specialist.

Piston Pin Bearings

Before you replace a piston, check for wear in the piston-pin bearing. The piston will be free to move from side to side on the pin, but if you detect any up-and-down movement in the bearing, you will have to replace the piston-pin bushing in the connecting rod.

Piston Pin Removal. The piston pin is normally held in place by a snap ring at each end. These are spring-tensioned rings that expand into a groove machined into the piston. In order to remove these rings, you need snap-ring pliers, which have hardened steel pins set in the end of each jaw. At each end of a snap ring is a small hole. Insert the tips of the pliers into the snap-ring holes and squeeze (see Figure 7-36).

If you don't have snap-ring pliers, you may be able to accomplish the job by grinding the jaws of a pair of needle-nose pliers to the proper size to fit the snap-ring holes (see Figure 7-37). Pinch the snap rings *only enough to remove them. Excessive squeezing and compression of snap rings is a major cause of failure later on.*

You only need to remove the snap ring from one end of the piston. If the piston pin does not slide out easily (some are an interference fit), dip the piston into near-boiling water to expand it before you tap out the pin. *Any attempt to force the pin out of a cold piston is likely to distort the piston permanently.* When the piston pin is out, check it for ridging, or burning due to oil starvation (in the latter case it will turn a blue color). If either is present to any marked degree, replace the pin.

Bushing Replacement. In order to push the old piston-pin bushing out of the connecting rod, you may also have to heat the rod in near-boiling water. Firmly support the connecting rod on blocks of wood, place the new bushing against the old one, hold a block of wood to the new bushing, and gently tap the block with a hammer. The new bushing will drive out the

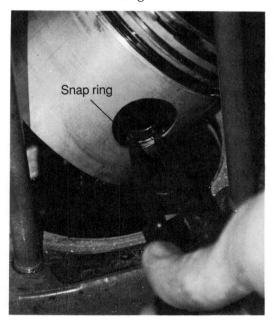

Figure 7-36. *Snap-ring pliers in action.*

Figure 7-37. *Use of needle nose pliers to remove a snap ring. (Courtesy Caterpillar Tractor Co.)*

old one. The procedure will go more easily if you cool the new bushing in a freezer for a few minutes before starting. Note that most bushings have an oil hole that *must* line up with an oil way in the connecting rod.

Now check the fit of the piston pin in the new bushing—many bushings require reaming before the pin fits. This is really work for a machine shop, but in the absence of the specialized tools, you can do it by wrapping a piece of wet-and-dry sandpaper (400 to 600 grit) around a suitably-sized piece of doweling, lubricating with diesel, and working back and forth. Be sure to hold the dowel square to the bushing, and to work around the whole bushing evenly. Do this until the piston pin will just slide into place: when the fit is correct, a lubricated piston pin will slide slowly through the bushing under its own weight.

Reassembly of a piston to its connecting rod is the reversal of disassembly, and may require the piston (but not the piston pin) to be heated. The pin can be put into a freezer or packed in ice, and will slide right into place.

Figure 7-38. *Aligning a bearing shell locking tab with its housing. (Courtesy Caterpillar Tractor Co.)*

Connecting Rod Bearings

Now is also the time to replace the connecting rod bearings, if this is necessary. These bearings consist of a precision-made steel shell lined with a special metal alloy (babbitt or lead bronze). You can remove the shells from the connecting rod and its cap by pushing on one end of each shell—they should slide around inside the rod or cap and slip out the other end (see Figure 7-38). A locating lug on the back of each shell at one end ensures that they can only be pushed out, and new ones inserted, in one direction.

If this is the engine's first major overhaul, it will almost certainly have standard-sized bearings. But an older engine may have had its crankshaft reground, in which case the crankshaft journals will be smaller than standard and the bearing shells will be correspondingly thicker. The back of the shells will be stamped STD, 0.010, 0.020, or 0.030, indicating the size of the new shells required.

If the engine has been knocking badly or the old bearings are seriously worn or scored (see Figure 7-39), the crankshaft journal may be damaged or worn into an ellipse (see Figure 7-40). The latter can only be measured with the appropriate micrometers, which will require a specialist's help. If the crankshaft is damaged or excessively worn, fitting new bearings is pointless because they will only last a short while.

The crankshaft will have to be removed and reconditioned, which is beyond the scope of this book. If the old bearings failed due to oil starvation, the crankshaft may have got excessively hot and will show heat discolorations. The heat may have destroyed the special hardening process that is applied to crankshaft bearing journals—a new crankshaft is needed.

When fitting new bearing shells to a connecting rod and its cap, the backs of the shells and the seating surfaces on the rod and cap must be spotlessly clean. You can push the new shells directly onto their seats or slide each around inside the housing until the lug seats in the slot (see Figure 7-41). The shells must seat squarely, and the lugs must be correctly positioned. *Make sure that any bearing shell that has an oil hole in it is fitted to the appropriate housing and lined up with its oilway.*

Replacing Pistons

When you place a piston into the cylinder from which it came, the crank for that cylinder should be at bottom dead center. Coat the cylinder, piston, and rings with oil, then gently lower the piston into the bore until the bottom ring, the oil-scraper ring, rests on the top of the cylinder (see Figure 7-42). At this point, arrange all the rings so that their end gaps are stag-

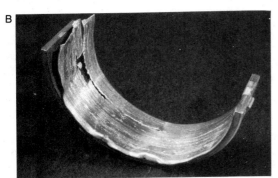

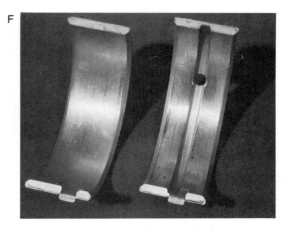

Figure 7-39. *(A) Extensive scratching from dirt in the oil. These shells cannot be used again. (B) This shell got so hot that it began to melt. (C) These shells are breaking up as a result of oil starvation. (D) This shell had a paint chip behind it due to improper cleaning at the time of installation. This resulted in severe localized overheating. The shell cannot be used again. (E) Another case of oil starvation. This crankshaft will have to be removed from the engine and reground. (F) A set of bearing shells in good condition—light scratching is normal, as long as the bearing journal on the crankshaft is shiny and smooth. (Courtesy Caterpillar Tractor Co.)*

A

C

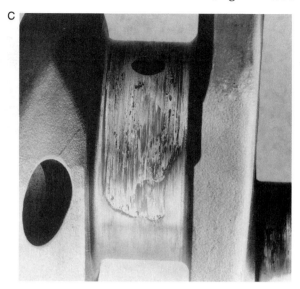

B

D

Figure 7-40. *(A) Heat distortion—discard. (B) Corrosion of crankshaft journal—cannot be used again. (C) Heat distortion, plus material from the bearing shell is adhering to the journal—cannot be used. (D) Cracked crankshaft—discard. (E) Smearing of bearing shell around the crankshaft—do not use again. (Courtesy Caterpillar Tractor Co.)*

E

Figure 7-41. *Aligning the oil hole in the bearing shell with its oilway. (Courtesy Caterpillar Tractor Co.)*

Figure 7-42. *Lowering a piston into the cylinder. (Courtesy Perkins Engines Ltd.)*

gered around the piston. This prevents blow-by through lined-up gaps.

Ring grooves on Detroit Diesels have a pin in them at one point. This is placed in the middle of the end gap and prevents the rings from turning on the piston, which might allow the ends of the rings to line up with the air-intake ports at the bottom of the cylinder. If this were to happen, the rings would try to spring out into the ports and would break.

Installing a Piston. You can rent a piston-ring clamp from an automotive parts store, if you know the piston's diameter. This holds the rings tightly in their grooves while you slide the piston into its cylinder (see Figure 7-43). You can dispense with the clamp, however, if you have to. You will need a helper and three screwdrivers, or similar blunt instruments (see Figure 7-44).

Push the ring on which the piston is sitting into its groove at its center point (the point opposite the ring gap) with one screwdriver. The helper then works around the ring in one direction, easing the ring into its groove until almost at one end. Hold the ring at this point with another screwdriver. The mechanic works around the ring in the other direction, easing it into its groove until almost at the other end, at which point it is held with the third screwdriver. The ring should now be all the way into its groove, and the mechanic

still has one hand free to tap the piston gently down into the cylinder, using the handle of a hammer or similar piece of wood (see Figure 7-44).

The piston will slide down until the next ring sits on top of the cylinder; repeat the procedure. Never use force—it will result in broken rings. Once all the rings are in the cylinder, push the piston down from above and guide the rod onto the crankshaft from below. Make sure the connecting rod is lined up squarely with the crankshaft journal. The crankshaft journal (bearing surface) must be spotlessly clean (use lint-free rags) and well-oiled. Replace the connecting rod cap.

It is a good practice to fit new cap bolts, whether they are called for or not. These bolts are subjected to very high loads, and if one fails, a tremendous amount of damage will result.

Torquing Procedures. When the cap nuts and bolts are replaced they must be tightened to a very specific torque. This will be given in the manufacturer's specifications (if not available, see Tables 7-1 and

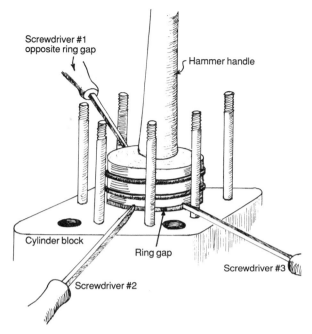

Figure 7-44. *Replacing a piston without a ring clamp.*

Figure 7-43. *Use of a piston-ring clamp. The clamp squeezes the piston rings into the grooves. (Courtesy Perkins Engines Ltd.)*

7-2 for general guidelines). You will need a *torque wrench.* These wrenches indicate exactly how much pressure is being applied to the nut or bolt.

Torque wrenches come in two basic types. You adjust the torque setting of the more expensive wrenches by screwing the end of the handle in and out, lining up a pointer with a scale on the body of the wrench. The scale indicates the torque pressure at which the wrench is now set. As you tighten the bolt or nut, the wrench will click when you reach this torque setting. You can hear and feel the click.

Cheaper wrenches have a flexible handle with a pointer attached to it, the end of which moves over a scale set across the wrench. As you apply pressure to a nut or bolt, the handle of the wrench flexes and the pointer moves across the scale. It is hard to use these wrenches with any degree of precision.

When you torque nuts or bolts, the threads must be clean, free-running, and generally oiled (some-

times a manual will specify a *dry* torque setting, without oil). Friction in the threads and sudden jerks on the wrench will give a false torque reading. The nuts or bolts must be pulled down with an even, steady pressure. On the most important nuts and bolts, over-tighten them by just a few pounds then back off and re-torque to the specified point. This ensures that everything is correctly pulled down. There should be some means of locking the nuts (a locking washer that bends over onto one of the flats on the nut, or the nut has a nylon insert).

At all critical bolt-tightening points in engine work, bolts must be tightened evenly. Just pinch one, and then do the bolt opposite; then apply a few more pounds of pressure to each one and continue in this fashion until the final torque setting is reached. Do this in a *minimum* of three stages.

When you replace bearing caps, the shaft they enclose should be turned a full revolution by hand after each increase in the tightening pressure until the full torque is reached. This ensures that there are no tight spots or binding. Tight spots must be removed; this procedure is especially important when fitting new bearing shells.

Table 7-1

Grade Identification Marking on Bolt Head	SAE Grade Designation	Nominal Size Diameter (inch)	Tensile Strength Min. (psi)
None	2	No. 6 thru 3/4 over 3/4 to 1 1/2	74,000 60,000
(mark)	5	No. 6 thru 1 over 1 to 1 1/2	120,000 105,000
(mark)	7	1/4 thru 1 1/2	133,000
(mark)	8	1/4 thru 1 1/2	150,000

Table 7-2

STANDARD BOLT AND NUT TORQUE SPECIFICATIONS

THREAD SIZE	GRADE 2 (lb-ft)	Nm	THREAD SIZE	GRADE 5 (lb-ft)	Nm
1/4-20	5- 7	7- 9	1/4-20	7- 9	10- 12
1/4-28	6- 8	8- 11	1/4-28	8- 10	11- 14
5/16-18	10- 13	14- 18	5/16-18	13- 17	18- 23
5/16-24	11- 14	15- 19	5/16-24	15- 19	20- 26
3/8-16	23- 26	31- 35	3/8-16	30- 35	41- 47
3/8-24	26- 29	35- 40	3/8-24	35- 39	47- 53
7/16-14	35- 38	47- 51	7/16-14	46- 50	62- 68
7/16-20	43- 46	58- 62	7/16-20	57- 61	77- 83
1/2-13	53- 56	72- 76	1/2-13	71- 75	96-102
1/2-20	62- 70	84- 95	1/2-20	83- 93	113-126
9/16-12	68- 75	92-102	9/16-12	90-100	122-136
9/16-18	80- 88	109-119	9/16-18	107-117	146-159
5/8-11	103-110	140-149	5/8-11	137-147	186-200
5/8-18	126-134	171-181	5/8-18	168-178	228-242
3/4-10	180-188	244-254	3/4-10	240-250	325-339
3/4-16	218-225	295-305	3/4-16	290-300	393-407
7/8- 9	308-315	417-427	7/8- 9	410-420	556-569
7/8-14	356-364	483-494	7/8-14	475-485	644-657
1- 8	435-443	590-600	1- 8	580-590	786-800
1-14	514-521	697-705	1-14	685-695	928-942

Grade identification markings are normally stamped on the heads of the bolts.

Tables 7-1 and 7-2. Adapted from tables supplied by Detroit Diesel Corp.

Replacing Cylinder Heads

The cylinder head and block must be spotlessly clean before you replace the head (see Figure 7-45). Any pieces of rag and so on used to block off the oil, water, and other passages in the block and head must be removed at this time. Set a new gasket on the block and lower the head onto it. Some gaskets have a top and will be appropriately labeled.

Although most manufacturers do not recommend it, metal gaskets often benefit from a little jointing

Figure 7-45. *A well-cleaned cylinder head face. (Courtesy Perkins Engines Ltd.)*

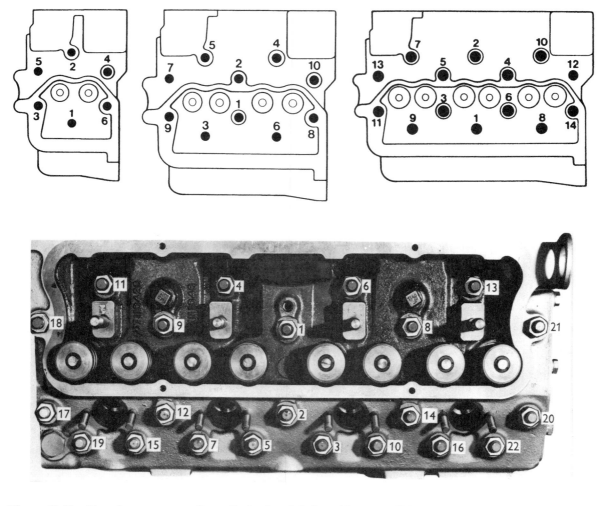

Figure 7-46. *Torquing sequence for cylinder-head bolts. (Courtesy Volvo Penta and Perkins Engines Ltd.)*

paste smeared on them, taking care not to get any down the passages. Jointing paste is available at automotive parts stores; be sure it is made for high-temperatures, is resistant to water and oil, and will withstand high pressure. Most fiber gaskets (increasingly the norm) are fitted without any paste. Always fit a new gasket if possible, even if the old one looks perfectly all right. It is tremendously aggravating to reassemble an engine with an old gasket only to find that the gasket leaks.

The head nuts (bolts) must be tightened evenly, a bit at a time as outlined above, until the manufacturer's specified torque setting is reached (see your owner's manual and Figure 7-46). Proper torquing procedures are more important here than on gasoline engines, owing to the much higher cylinder pressures generated by diesel engines.

If you don't follow the correct bolt-tightening sequence, uneven pressure may develop and lead to a blown head gasket or a warped cylinder head. If the manufacturer's recommended torque sequence is not available, you can safely assume that the center nuts are pulled down first. From then on, work out to the ends of the cylinder head, tightening a nut on one side of the center to a particular torque setting, and then one on the other side, and so on until all are done. Then increase the torque setting and do this again. Once again, torquing should be done in a *minimum* of three stages.

The correct torque setting is especially critical on engines with dissimilar metals (e.g., a cast-iron block and an aluminum cylinder head), because the metals have differing rates of expansion and contraction. The correct torque setting is more important on engines with a greater number of cylinders (e.g., six as opposed to four). Always recheck the torque settings after an engine has been reassembled and run for a while. A head gasket, especially a metal one, will occasionally settle, loosening the head nuts and creating the potential for a blown gasket.

Replacing Push Rods and Rockers

Before you replace the push rods, roll them on a flat surface to make sure they are straight. A chart table or galley countertop will be level enough; the bed of a table saw or drill press is even better. Any bend will

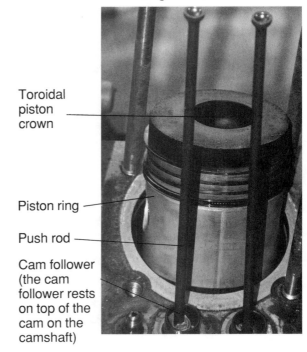

Toroidal piston crown

Piston ring

Push rod

Cam follower (the cam follower rests on top of the cam on the camshaft)

Figure 7-47. Push rods and cam followers. (The cylinder has been removed and the push rods have been wedged in place with paper to illustrate their location.)

be immediately apparent.

Place the push rods in their respective holes in the cylinder head and block (round ends down, cupped ends up) and be sure each is properly seated in a hollow in what is known as a *cam follower* (see Figure 7-47). You can't see this seat, but if you've missed it the push rod will be cockeyed and will probably be resting on the rim of the cam follower. In some engines, the push rods share a common space, which makes it possible to miss the cam follower altogether or even to hit the wrong one. In most engines this cannot be done. If the push rod is centered in its hole in the cylinder head and feels firmly cupped at its lower end, it is seated correctly. Note that some push rods will be sticking up more than others.

The rockers go on next, but before fitting them, the lock nut at the end of each rocker arm should be loosened and the screw it locks undone a couple of turns (see Figure 7-48). This is just a little safety pre-

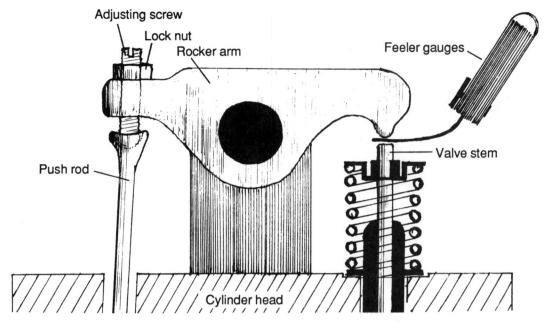

Figure 7-48. *Setting valve clearances.*

caution in case the valve timing has been upset or valve clearances radically changed. It prevents any risk of forcing a valve down onto the piston crown and bending the valve stem when the rocker bolts are tightened up. Torque down the rocker assembly to the manufacturer's setting.

Retiming an Engine

Decarbonizing will not disturb the timing of an engine with push rods, but any time an overhead camshaft is removed, valve timing is upset and will need to be reset. Since the fuel injection pump is tied in with the valve timing on all engines, the following procedure for retiming an overhead camshaft coincidentally describes how to retime a fuel injection pump on any engine.

General Principles. Engine timing involves the timing drive gear, which is keyed to the end of the crankshaft; the camshaft drive gear, which operates the valve timing; and the fuel-injection-pump drive gear. On 2-cycle engines, these gears are all the same size because the camshaft and fuel injection pump rotate at the same speed as the engine. On 4-cycle engines, the timing gear is half the size of the other two because the crankshaft must rotate twice for every complete engine cycle.

Engine timing is set by getting these three gears in exactly the correct relationship to one another. Each of the gears involved in engine timing has a punch mark or line somewhere on its face. When engine timing is belt-driven or chain-driven, these marks are lined up with corresponding marks on the timing-gear housing. When engine timing is transmitted through intermediate gears, the intermediate gears have punch marks that line up with the marks on the timing gears (see Figure 7-49). This alignment should be exact—if it is not, something is wrong.

Specific Procedures. Timing is always done at top dead center (TDC) on the compression stroke of the number 1 cylinder (the one at the front end, or timing-gear end, of the engine). When the engine is at TDC on the number 1 cylinder, the keyway in the end of the crankshaft, which positions the timing drive gear, will also be at TDC.

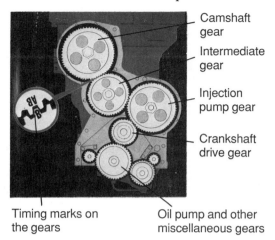

Camshaft gear

Intermediate gear

Injection pump gear

Crankshaft drive gear

Timing marks on the gears

Oil pump and other miscellaneous gears

Figure 7-49. *Timing marks. (Courtesy Caterpillar Tractor Co.)*

Two-cycle engines have only one TDC because the engine fires on every revolution of the crankshaft, but 4-cycle engines have two TDCs—one on the compression stroke and one on the exhaust stroke. You time the engine at TDC on the compression stroke. Normally you can determine which stroke it is by looking at the position of the valves, but with the camshaft off this is not possible.

If the fuel injection pump timing has not been disturbed, the mark on its drive gear will line up with a corresponding mark on the timing gear housing, or on an intermediate gear, when the engine is at TDC on the compression stroke of the number 1 cylinder. This will tell you that you have the correct TDC.

If the fuel pump timing has been disturbed, then it does not matter at which TDC on the number 1 cylinder you set the timing, because the camshaft and fuel pump will be timed together. You must, however, time the fuel injection pump and the camshaft at the same TDC—which one is immaterial. Otherwise it would be possible to have the injection pump injecting the cylinders when the pistons were at the top of their exhaust stroke; the engine would never run. This is known as the timing being out 180°—although the engine is a full revolution out (360°), the camshaft and injection pump turn at half engine speed on 4-cycle engines, so one of them would be out 180°.

With the engine at TDC on the number 1 cylinder

and the crankshaft keyway also at TDC, line up all the gear marks with their corresponding marks on the gear housing or intermediate gears, then install the belt, chain, or intermediate gears (see Figure 7-50). Always double-check to see that all the marks still line up after you tension the belt or chain—sometimes one of the gears will move around by one tooth, in which case timing will have to be repeated. If the timing is belt-driven, I recommend fitting a new belt. If chain driven, the chain's length, stretched, should be compared to the manufacturer's specified length to see if it needs replacing.

This completes basic timing. All that remains to do is to fine-tune the fuel injection pump. In almost all instances, the injection pump will be bolted to the other side of the timing gear housing. (On occasion it is bolted to a little platform of its own.) The flange on the pump that bolts up to the gear housing has machined slots for its bolts, which means that even after the gear timing has been set, the pump can still be rotated to the extent allowed by the slots. This rotation does not move the timing gear but turns the pump around its drive shaft.

A line scribed on the pump flange and the timing gear housing *must be exactly lined up before the pump flange is tightened* (see Figure 7-51). This completes the injection pump timing.

Detroit Diesel injection timing requires a special timing gauge and specific instructions (see Figure 7-52). It is straightforward enough, but you will need the relevant shop manual and gauge.

Valve Clearances, 4-Cycle Engines

All valves have a small clearance between the valve stem and rocker arm when they're fully closed. It is important to maintain the manufacturer's specified clearance. Too little clearance causes a valve to stay slightly open at all times, as the engine heats up and the metal parts expand. This results in lost compression and a burned seat and valve. If the clearance is too great, the valve will open slightly late, won't open far enough, and will close a little too soon.

General Principles. The inlet valve of a 4-cycle engine opens on the downward stroke of its piston. Both valves are closed during the next upward (com-

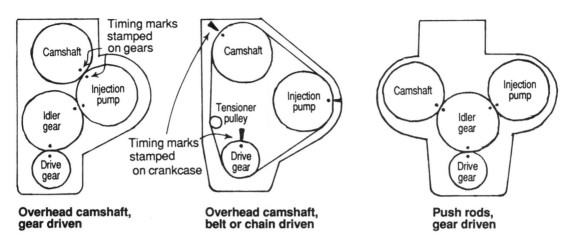

Overhead camshaft, gear driven

Overhead camshaft, belt or chain driven

Push rods, gear driven

Figure 7-50. *Timing arrangements.*

Figure 7-51. *Fuel injection pump (CAV type DPA) mounted vertically. (1) Timing marks scribed on the pump's mounting flange and engine's timing cover. (2) Idle-speed adjusting screw. (3) Maximum-speed screw. (Do not tamper with the maximum-speed screw. Although you cannot see it in this photo, the screw has a seal on it, and breaking it automatically voids the engine's warranty.) (Courtesy Perkins Engines Ltd.)*

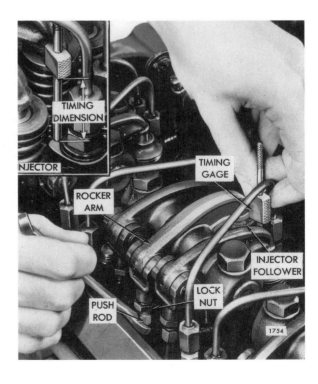

Figure 7-52. *Timing the ignition of a Detroit Diesel, using a special gauge. (Courtesy Detroit Diesel Corp.)*

pression) stroke, and for most of the following (power) stroke. The exhaust valve then opens and remains open on the next upward (exhaust) stroke. At the top of this stroke the exhaust valve is closing at the same time the inlet valve is opening. This is known as *valve overlap;* the valves are *rocking.* By watching the movement of the rocker arms while you slowly rotate the engine, you can determine where in the cycle each piston is, therefore the position of the cam that operates each push rod.

You set the valve clearance when the valve is fully closed at TDC on the compression stroke. On engines with overhead camshafts you can see the cams, and valve clearances should be set when a cam is 180° away from the rocker it operates. On engines with push rods, where the operation of the camshaft cannot

be observed, the following method will establish the correct point for setting valve clearances.

Specific Procedures. In order to find top dead center for any cylinder, slowly rotate the crankshaft in its normal direction of rotation. Watch the inlet valve's push rod as it moves up and down. When it is almost all the way down, the piston is at the bottom of its inlet stroke. Mark the crankshaft pulley, and turn the engine another half a revolution. Now the piston will be close to TDC on its compression stroke and you can set the valve clearances on this cylinder. On most engines the crankshaft pulley is marked for TDC on the number 1 cylinder, but on older engines you should not rely on any mark—someone may have changed things around at some time.

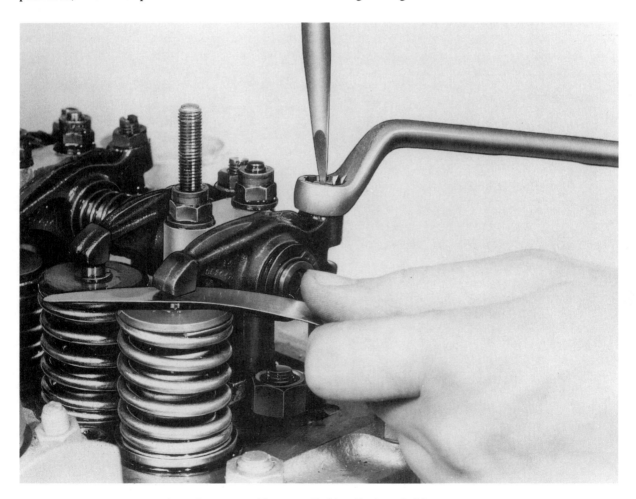

Figure 7-53. *Setting a valve clearance. (Courtesy Perkins Engines Ltd.)*

The manufacturer's specifications will indicate valve clearances in millimeters or thousandths of an inch, and whether the valves should be adjusted hot or cold. If they are to be set when the engine is hot, an initial adjustment will have to be made cold then checked again after the engine has been run. Place an appropriate feeler gauge between the top of the valve stem and the rocker arm (see Figure 7-53). Tighten the adjusting screw until the arm just begins to pinch the feeler gauge. Now, tighten the lock nut on the adjusting screw and check the clearance again in case something slipped. This valve is set; adjust the other one on the same cylinder. (Note: Although most valves are adjusted this way, some engines have threaded push rods, which are screwed in and out to adjust valve clearances, whereas other engines have

# OF CYLINDERS	VALVES "ROCKING" ON:	SET VALVE CLEARANCE ON:
4	4	1
	2	3
	1	4
	3	2
6	6	1
	2	5
	4	3
	1	6
	5	2
	3	4

Figure 7-54. *Sequence for setting valve clearances on 4-cycle engines (not 2-cycle).*

an adjusting nut in the center of the rocker assembly.)

On a two-cylinder engine, when one piston is at TDC the other is at BDC. After the valves on the first cylinder are set, a half-turn will bring the piston on the other cylinder to TDC. A quick glance at the valves will show if this is TDC on the exhaust stroke (the valves will be rocking) or the compression stroke. If it is the former, you will have to rotate the engine another full turn. The valve clearances on the second cylinder can now be set.

On three-cylinder and six-cylinder *in-line* engines, one-third of a revolution will always bring another piston to TDC. The pistons of six-cylinder in-line engines move in pairs—normally numbers one and six together, two and five, and three and four (see Figure 7-54). When one piston of a pair is at TDC on its exhaust stroke (valves rocking), the other piston in the pair will be at TDC on its compression stroke and you can set its valves.

The pistons of four-cylinder in-line engines also move in pairs—one and four together, and two and three—with a half turn separating TDC between the pairs. When the valves on either one of a pair of pistons are rocking, the other piston in the pair is at TDC on its compression stroke and its valves can be set.

If you are at all unsure of how this works, write down all the cylinders on a piece of paper, showing which ones operate together. The inlet and exhaust valves should be determined from their respective manifolds and the engine turned over slowly a few times to familiarize yourself with the valve opening and closing sequence. Careful attention to the logic of the situation will soon indicate where the pistons are and when to set valve clearances. Be sure to turn the engine over in its normal direction of rotation.

Valve Clearances, 2-Cycle Engines

Most 2-cycle diesels have no inlet valves but instead have two or more exhaust valves per cylinder. Remember that these valves open when the piston is near the bottom of the power stroke and close when the piston is partway up the compression stroke. From the point of closure, another one-third of a turn brings the piston more or less to TDC. At this time, the rocker arm operating the injector for this cylinder will be fully depressed (i.e. on its injection stroke). The valve clearances can now be set.

Accessory Equipment

The final step in decarbonizing is to refit all the fuel lines, manifolds, valve cover, turbocharger, and anything else that was removed. If no other part of the fuel system has been broken loose, a few turns of the engine should push diesel up to the injectors. Otherwise the fuel system will need bleeding as explained on page 58.

Troubleshooting Chart 7-1.

Cause	Seizure	Excessive oil consumption	Rising oil level	Low oil pressure	High exhaust back pressure	Loss of power	Overheating	Poor idle	Hunting	Misfiring	White smoke	Blue smoke	Black smoke	Knocks	Low cranking speed	Lack of fuel	Low compression	Poor starting
Throttle closed/fuel shut-off solenoid faulty/tank empty																●		●
Lift pump diaphragm holed						●										●		●
Plugged fuel filters						●		●		●						●		●
Air in fuel lines						●		●		●	●		●	●		●		●
Dirty fuel						●		●		●			●	●				●
Defective injectors/poor quality fuel						●		●		●			●	●				●
Injection pump leaking by						●	●	●		●			●	●				●
Injection timing advanced or delayed							●							●				
Too much fuel injected		●				●		●		●	●	●	●					
Piston blow-by						●		●		●	●						●	●
Dry cylinder walls						●		●		●	●						●	●
Valve blow-by						●	●	●		●	●						●	●
Worn valve stems												●						
Decompressor levers on/valve clearances wrong/valves sticking						●		●		●	●						●	●
Pre-heat device inoperative								●		●								●
Plugged air filter						●							●					●
Plugged exhaust/turbocharger/kink in exhaust hose					●	●	●											
Oil level low	●			●			●											
Wrong viscosity oil	●	●		●											●			
Diesel dilution of oil	●		●	●														
Dirt in oil pressure relief valve/defective pressure gauge				●														
Governor sticking/loose linkage								●	●									
Governor idle spring too slack								●										
Defective water pump/defective pump valves/air bound water lines	●						●											
Closed sea cock/plugged raw-water filter or screen/plugged cooling system	●						●											
Blown head gasket/cracked head/water in cylinders	●	●					●			●	●						●	●
Uneven load on cylinders							●										●	●
Worn bearings				●			●											●
Seized piston				●		●												●
Auxiliary equipment engaged															●			●
Battery low/loose connections															●			●
Engine overload/rope in propeller	●					●	●						●					

Chapter 8

Marine Transmissions

Manual transmissions (gearboxes) are still found in some boats, and they are generally of the planetary, or epicyclic, type. Far more common are hydraulic planetary boxes, many of which are made by Borg-Warner, and servo-hydraulically operated two-shaft boxes, many of which are made by Hurth, with Borg-Warner now producing a very similar transmission. The picture is complicated a little further by a new addition from Borg-Warner that combines the principles of a two-shaft transmission with the hydraulic clutch packs found in hydraulic planetary transmissions—I am going to refer to this as a "hybrid two-shaft transmission." Borg-Warner and Hurth have a major share of the worldwide market in small marine transmissions. Other makes are very similar to these two.

Planetary Transmissions

In this type (manual and hydraulic), the engine turns a geared drive shaft (the *drive* gear), which rotates constantly in the same direction as the engine. Deployed around and meshed with this gear are two or three gears (the *first intermediate* gears) on a carrier assembly. The first intermediate gears mesh with more gears (the *second intermediate* gears), also mounted on the carrier assembly. The second intermediate gears engage a large geared outer hub. The carrier assembly, with its collection of first and second intermediate gears, is keyed to the output shaft of the transmission (see Figure 8-1).

On one end of the drive shaft is the forward clutch. Engaging forward locks the entire drive shaft and carrier assembly together—the whole unit rotates as one, imparting engine rotation to the output shaft of the transmission via the carrier assembly (see Figure 8-2).

Reverse is a little more complicated. The forward clutch is released and a second clutch engaged. This locks the outer geared hub in a stationary position. Meanwhile the drive gear is still rotating the intermediate gears. Unable to spin the outer hub, the intermediate gears rotate around the inside of the hub in the *opposite* direction of the drive gear. The carrier assembly imparts this reverse motion to the output shaft (see Figure 8-3).

Manual and hydraulic versions of a planetary box are very similar. The principal difference is that the reverse clutch of a manual box consists of a brake band that is clamped around the hub (see Figure 8-4), whereas in a hydraulic box a second clutch, similar to the forward clutch, is used.

A manual box uses pressure from the gearshift

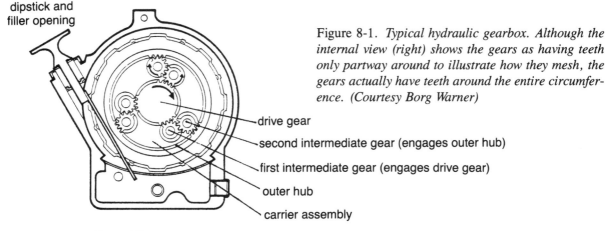

dipstick and filler opening

drive gear

second intermediate gear (engages outer hub)

first intermediate gear (engages drive gear)

outer hub

carrier assembly

Figure 8-1. *Typical hydraulic gearbox. Although the internal view (right) shows the gears as having teeth only partway around to illustrate how they mesh, the gears actually have teeth around the entire circumference. (Courtesy Borg Warner)*

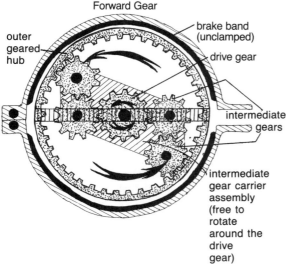

Forward Gear

outer geared hub

brake band (unclamped)

drive gear

intermediate gears

intermediate gear carrier assembly (free to rotate around the drive gear)

Figure 8-2. *Forward in a planetary gearbox. The brake band is unclamped, and the forward clutch locks together the drive gear and carrier assembly so that they rotate as one. (The band across the center of the drawing symbolizes the clutch locking up all the gears.)*

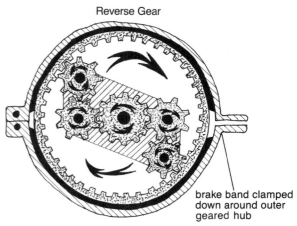

Reverse Gear

brake band clamped down around outer geared hub

Figure 8-3. *In reverse, the forward clutch is released, leaving the carrier assembly free to rotate around the drive gear while the brake band is clamped down, locking the geared hub. The carrier assembly is driven around the hub in the opposite direction of the drive gear.*

lever to engage and disengage the clutches. A hydraulic box incorporates an oil pump. The gearshift lever merely directs oil flow to one or the other clutch; the oil pressure does the actual work. While quite a bit of pressure is needed to operate a manual transmission, gear shifting with a hydraulic transmission is a fingertip affair.

Two-Shaft Transmissions

In this type, the engine is coupled to the input shaft, to which two gears are keyed, one at either end. A second shaft, the output shaft, has two more gears riding on it; one of these engages one of the input gears directly, and the other engages the second input gear via an intermediate gear (see Figure 8-5). These two output gears are mounted on bearings and freely rotate around the output shaft.

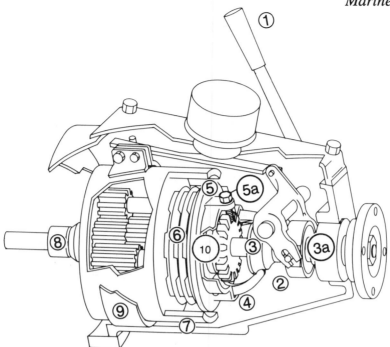

Figure 8-4. *Operating principles of a Paragon S-A-O-a common planetary gearbox. To engage forward: (1) push lever; (2) shift yoke pushes back; (3) shift cone slides along output shaft (3A), thus moving (4) the cam levers. The cam levers press on (5) the pressure plate and compresses the friction discs (6) in the clutch. The friction discs press against the near surface of (7) the gear carrier, completing the transmission of motion from the (8) input shaft to the output shaft (3A). Adjust the pressure plate by backing out the lock bolt (5A), screwing up the castellated nut (10) and retightening the lock bolt. To reverse direction, pull back lever to compress (9) the reverse band, locking it around the gear carrier like a huge hose clamp.*

The drive gears impart continuous forward and reverse rotation to the output gears. Each output gear has its own clutch, and between the two clutches is an engaging mechanism. Moving the engaging mechanism one way locks one gear to the output shaft, giving forward rotation; moving the mechanism the other way locks the other gear to the shaft, giving reverse rotation (see Figure 8-5).

When the clutch engaging mechanism is first moved to either forward or reverse, it gently presses on the relevant clutch. This initial friction spins a *disc carrier*, which holds some steel balls in tapered grooves. The rotation drives the balls up the grooves. Because of the taper in the grooves, the balls exert an increasing pressure on the clutch, completing the

engagement (see Figure 8-6). Only minimal pressure is needed to set things in motion, and thereafter a clever design supplies the requisite pressure to make the clutch work, but without the necessity for oil pumps or oil circuits. Gear shifting is once again a fingertip affair.

Hybrid Two-Shaft Transmissions. The principles are similar to a two-shaft transmission, although the box geometry may be varied somewhat to incorporate a down angle on the output shaft (see Figure 8-7). However, for gear shifting, instead of using the servo-hydraulic principles of a standard two-shaft transmission, the hybrid transmissions incorporate an oil pump that provides the pressure for operating the clutches, just as in a hydraulic planetary transmission.

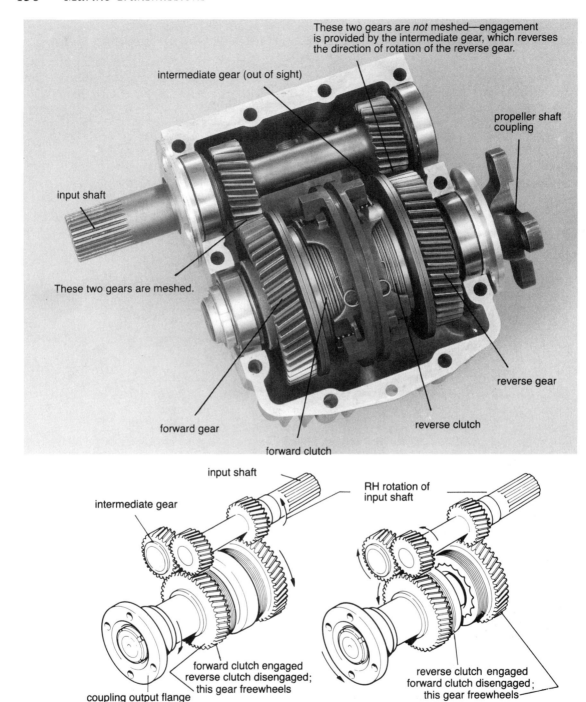

These two gears are *not* meshed—engagement
is provided by the intermediate gear, which reverses
the direction of rotation of the reverse gear.

intermediate gear (out of sight)

propeller shaft
coupling

input shaft

These two gears are meshed.

reverse gear

forward gear

reverse clutch

forward clutch

input shaft

RH rotation of
input shaft

intermediate gear

forward clutch engaged
reverse clutch disengaged;
this gear freewheels

reverse clutch engaged
forward clutch disengaged;
this gear freewheels

coupling output flange

Figure 8-5. *Two-shaft transmission.   With the forward clutch engaged (left), the output
shaft rotates in the direction opposite to the input shaft. With the reverse clutch engaged
(right), the input shaft drives the output shaft in the same direction via the intermediate
gear. (Courtesy Hurth)*

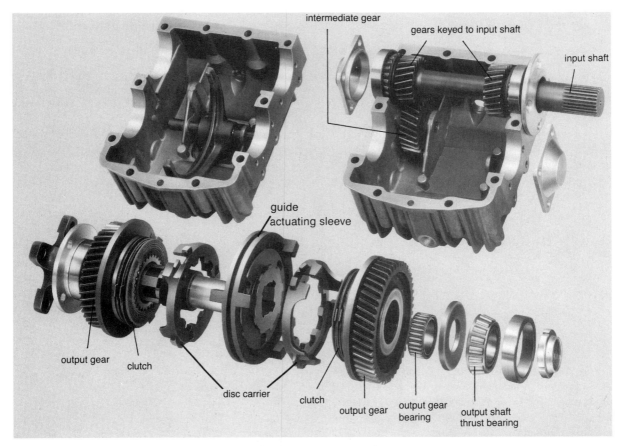

Figure 8-6. *The same transmission, disassembled. (Courtesy Hurth)*

Shaft Brakes

When a boat is being towed or is under sail with the motor shut down, the flow of water over a fixed-blade propeller (for more on propellers, see Chapter 9) will spin the propeller unless the propeller shaft is locked in some way. This is of little concern with manual and two-shaft transmissions, except that it creates unnecessary wear on bearings, oil seals, and the stuffing box. But *on some hydraulic transmissions (e.g., some Detroit Diesels), it will lead to a complete failure of the transmission* since no oil is pumped to the bearings when the engine isn't running. Some output-shaft oil seals (particularly rawhide seals) also will fail.

Boats with folding and feathering propellers do not need a shaft brake. You can brake a manual trans-mission simply by putting it into gear (reverse is best), but since a hydraulic transmission has no oil pressure when it's not operating, there is no way to put it into gear. A separate shaft brake is needed. Some are hydraulically operated, others are manual (see Figure 8-8). Hydraulic units have a spring-loaded piston, which is opposed by oil pressure from the transmission circuit when the engine is running. The oil pressure forces back the piston against its spring, releasing the brake (some form of a clamp around the propeller shaft or coupling). When the engine is shut down the spring forces the piston out, engaging the brake. Manual units are similar, but must be set and released by hand.

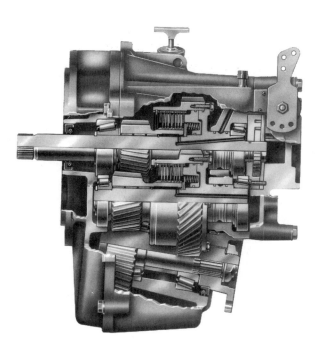

Figure 8-7. *A hybrid two-shaft transmission. This cutaway view shows the equal-size clutch packs that enable the unit to be operated in either forward or reverse. Note the transmission's helical gearing and tapered bearings. (Courtesy Borg Warner)*

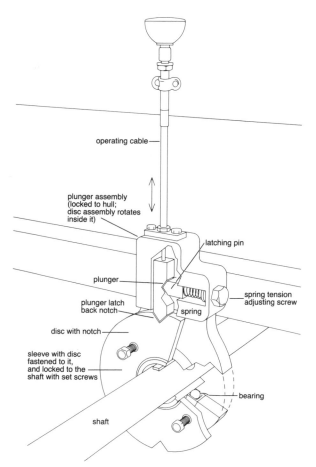

Figure 8-8. *The anatomy of a manual shaft lock. (Courtesy Shaft Lok)*

Maintenance

Transmission maintenance is minimal. It generally boils down to keeping the exterior clean (important for detecting oil leaks); periodically checking the oil level (see Figure 8-9) unless there is a leak, it should never need topping up; checking for signs of water contamination (water emulsified in oil gives it a creamy texture and color); and changing the oil annually. If the transmission has an oil cooler, especially a raw-water-cooled oil cooler, you must check the sacrificial zinc anodes regularly and change them when they are only partly eaten away (see the relevant sections on engine oil coolers).

Most transmissions operate on 30-weight engine oil or F-type transmission fluid. The latter is prefer-

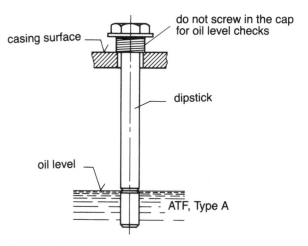

Figure 8-9. *Checking oil level on Hurth gearboxes. (Courtesy Hurth)*

red by many manufacturers (see your manual). If you have a hydraulic transmission, run the engine for a couple of minutes then shut it down before checking the oil level. If there is an oil screen, a magnetic plug, or both in the base of the transmission, inspect them for any signs of metal particles or other internal damage when changing the oil.

Clutch Adjustments on Manual Transmissions

The top of the box generally unbolts and lifts off. Inside are adjustments for forward and reverse gears. Reverse is easier.

Move the gear lever into and out of reverse—the brake band will be clearly visible as it clamps down on the hub and unclamps (see Figure 8-4). On one side of the band will be an adjusting bolt. If the transmission is slipping in reverse, tighten the bolt a little at a time, engaging reverse between each adjustment. When the gear lever requires firm pressure to go into gear, and clicks in with a nice, clean feel, adjustment is correct.

It is important not to overdo things. Put the box in neutral and spin the propeller shaft by hand. *If the brake band is dragging on the hub, it is too tight: the box is going to heat up and wear will be seriously accelerated.* If no amount of adjustment produces a clean, crisp engagement, the brake band is worn out and needs replacing—or at least relining.

To adjust the forward clutch, first put the transmission into and out of gear a few times to see what is going on. The main plate on the back of the clutch unit (it pushes everything together) will have either one central adjusting nut, or between three and six adjusting nuts around it. Tighten the central nut by one flat. For multiple adjusting nuts, put the box into neutral and turn it over by hand, tightening each adjustment nut as it becomes accessible by one-sixth of a turn. After going all the way around, try engaging the gear again. Repeat until the gear lever goes in firmly and cleanly. Lock the adjusting nuts. Do not tighten to the point at which the clutch drags in neutral; if overtightening seems necessary, the friction pads on the clutch plates probably are worn out and need replacing.

Clutches on Hydraulic and Two-Shaft Transmissions

These clutches are not adjustable. *The majority of problems that develop arise as a result of problems with the control cables, rather than the transmissions themselves* (see below—Hurth clutches, in particular, are very sensitive to improper cable adjustments). However there are a couple of problems peculiar to hydraulic boxes that need to be noted.

A buzzing sound indicates air in the hydraulic circuit, generally as a result of a low oil level. This will lead to a loss of pressure and slipping clutches. Most hydraulic transmissions also have an oil pressure regulating valve that passes oil back to the suction side of the oil pump if excess pressure develops. If the valve sticks in the open position, the clutches may slip or not engage at all. If, on the other hand, it sticks closed, the clutches will engage roughly (as they also will do if gear shifting is done at too high an engine speed). The valve is generally a spring-loaded ball or piston screwed into the side of the transmission— removing it for inspection and cleaning is easy (see Figure 8-10).

Overheating

Heavily loaded transmissions, especially hydraulic transmissions, tend to get hot (too hot to touch). In fact, many that do not have an oil cooler would benefit from the addition of one. Excessively high temperatures, however, are only likely to arise if the oil level is low (a smaller quantity of oil has to dissipate the heat generated), if the clutches are slipping (creating excessive friction), or if the oil cooler is not operating properly.

A slipping clutch should be evident from a loss of performance. The intense heat generated will soon warp clutch plates and burn out clutch discs. The oil in the transmission will take on a characteristic black look.

Oil cooler problems may arise on the water and oil sides. Transmission oil generally remains pretty clean, but a slipping clutch and other problems occasionally form a sludge that can plug up the oil side of the cooler. More likely is silt, corrosion, and scale

Troubleshooting Chart 8-1.

Transmission Problems.

Symptoms: Failure to engage forward or reverse; clutch drag in neutral; or tendency to stick in one gear.

Move the remote control lever through its full range a couple of times. Does it move the operating lever on the transmission itself through its full range? **YES** ↓	**NO** ▶ Check for a broken, disconnected, slipping, or kinked cable.
Is the remote control lever free-moving? **YES** ↓	**NO** ▶ Break the cable loose at the transmission and try again. If still stiff, remove the cable from its conduit, clean, grease, and replace. If the cable moves freely when disconnected from the transmission, move the transmission lever itself through its full range. If binding, the transmission needs professional attention.
When the remote control is placed in neutral, is the transmission lever in neutral? **YES** ↓	**NO** ▶ Adjust the cable length.
Is the transmission oil level correct? (Most transmissions have a dipstick; hydraulic transmissions frequently make a buzzing noise when low on oil). **YES** ↓	**NO** ▶ Add oil and run the engine in neutral to clear out any air.
Does the transmission output coupling turn when the transmission is placed in gear? **YES** ↓	**NO** ▶ The transmission needs professional attention.
Does the propeller shaft turn when the transmission coupling turns? **YES** ↓	**NO** ▶ The coupling bolts are sheared or the coupling is slipping on the propeller shaft. Tighten or replace set screws, keys, pins and coupling bolts as necessary.

There must be a fault with the propeller:
1. It may be missing or damaged;
2. A folding propeller may be jammed shut;
3. A variable pitch propeller may be in the "no pitch" position;
4. If this is the first trial of a propeller, it may simply be too small and/or have insufficient pitch.

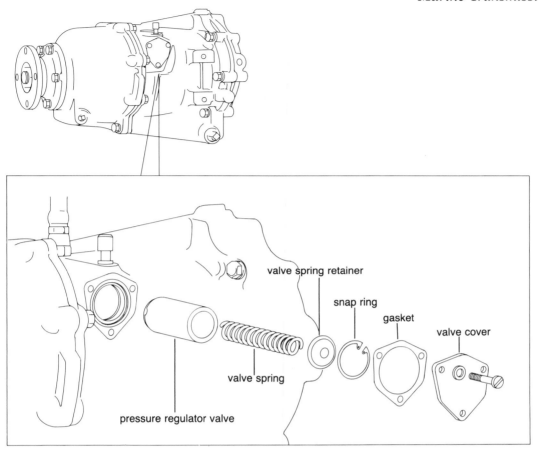

Figure 8-10. *Oil pressure regulating valve.*

interfering with the heat transference on the water side, especially if the cooler is raw-water cooled. (For cleaning and descaling see page 93).

Water in the Transmission

If the transmission has an oil cooler, this is the most likely source of water, especially if the cooler is the raw-water type. Pinholes form in cooler tubes just as in engine-oil coolers. Regular inspection and changing of sacrificial zinc anodes is essential. The only other likely source of water ingress is through the transmission output seal. For this to happen, the seal must be seriously defective, and the bilges must have large amounts of water slopping around, both of which were far more common years ago, when leather seals and wooden boats were the norm, than now.

Loss of the Transmission Oil

The rupture of an external oil line will produce a sudden, major, and catastrophic loss of oil, which will be immediately obvious. Less obvious will be the loss of oil through a corroded oil cooler. If it is raw-water cooled, the oil will go overboard to form a slick; if freshwater cooled, it will rise to the top of the header tank (see page 81).

Although the seal around the clutch-actuating lever, or the seal on a hydraulically operated shaft

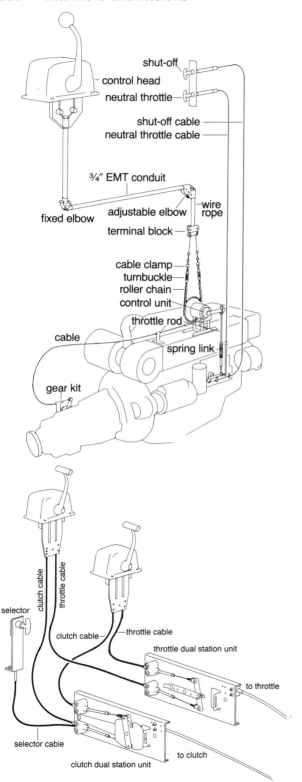

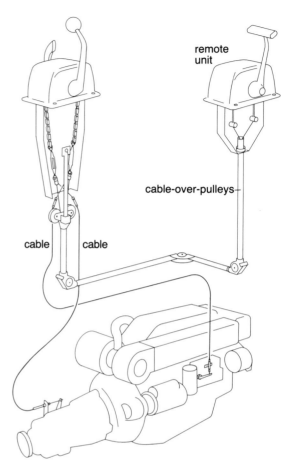

Figure 8-11. *(Top left) Typical remote engine and transmission controls. These can comprise three sections: (1) the pilothouse control; (2) the cable system; and (3) the engine control unit. Input motion is transmitted from the pilothouse control via the cables to the actuating mechanisms on the engine and transmission. This setup uses Morse's enclosed cable-over-pulley system. (Top right) A dual-station installation, with the main station controlled by push-pull cables, and the remote station using cable-over-pulley. (Bottom left) A dual-station installation using remote bellcrank units and single-lever controls. (Courtesy Morse Controls)*

brake, occasionally leaks small quantities of oil, the most likely candidate for this kind of leak is the output-shaft oil seal. This is true particularly if the engine and propeller shaft are poorly aligned (leads to excessive vibration), or if the propeller shaft has been allowed to freewheel when the boat is under sail. Alignment checks are covered on page 184. Seal replacement is dealt with below. On rare occasions, since the oil pressure in hydraulic transmissions is much higher than that in engines, the oil gets pumped through some ruptured seal into the flywheel housing or engine crankcase.

Control Cables

With the exception of manual boxes, which have a gear shift handle, most transmissions today use a push-pull cable to move the shift lever on the box (see Figure 8-11). A push-pull cable is one that pushes the lever in one direction, and pulls it in the other. *More transmission problems are caused by cable malfunctions than anything else. Faced with difficulties, always suspect the cable before blaming the box.*

If the transmission operates stiffly, fails to go into either or both gears, stays in one gear, or slowly turns the propeller when in neutral (*clutch drag*), make the following checks:

- See that the transmission actuating lever (on the side of the transmission) is in the neutral position when the remote control lever (in the cockpit or wheelhouse) is in neutral.
- Ensure that the actuating lever is moving fully forward and backward when you put the remote control into forward and reverse. This is particularly important on Hurth boxes.
- Disconnect the cable at the transmission and double-check that the actuating lever on the box is *clicking* into forward, neutral, and reverse. Note: Hurth boxes do not have a distinct *click* when a clutch is engaged. However, the lever must move through a minimum arc of 30 degrees in either direction. Less movement will cause the clutches to slip. More is OK. As the clutches wear, the lever must be free to travel farther. If the transmission actuating lever is stiff, or not traveling far enough in either direction, make sure that it is not rubbing on the transmission housing or snagging any bolt heads.
- While the cable is loose, operate the remote control to see if the cable is stiff. If so, replace the cable. Note that forcing cables through too tight a radius when routing them from the control console to the engine is a frequent cause of problems (see below).
- Inspect the whole cable annually, checking for the following (see Figure 8-12): Seizure of the swivel at the transmission end of the cable conduit; bending of actuating rods; corrosion of the end fittings at either end; cracks or cuts in the conduit jacket; burned or melted spots; excessively tight curves or kinks (the minimum radius of any bend should be 8 inches); separation of the conduit jacket from its end fittings; or corrosion under the jacket (it will swell up). If at all possible, remove the inner cable and grease it with a Teflon-based waterproof grease before replacing. Replace cables at least every five years and keep an old one as a spare.

Replacing an Output Shaft Seal

Unbolt and separate the two halves of the propeller coupling (see Figure 8-13). Mark both halves so that they can be bolted back together in the same relationship to one another.

On some boats with vertical rudderposts, the propeller shaft cannot be pushed far enough aft to provide the necessary room to slide the transmission coupling off its shaft. The propeller hits the rudderstock and will go no farther. In this case, the rudder has to be removed or the engine lifted off its mounts to provide the necessary space—an awful lot of work to change an oil seal. In such a case, you may want to consider having the propeller shaft shortened and installing a small stub shaft in-line between the transmission and propeller shaft.

The coupling half attached to the transmission output shaft must be removed. This coupling is held in place with a central nut, which is done up tightly on most modern boxes, while on some older boxes it is just pinched up and then locked in place with a cotter pin (split pin).

The coupling rides on either a splined shaft (one

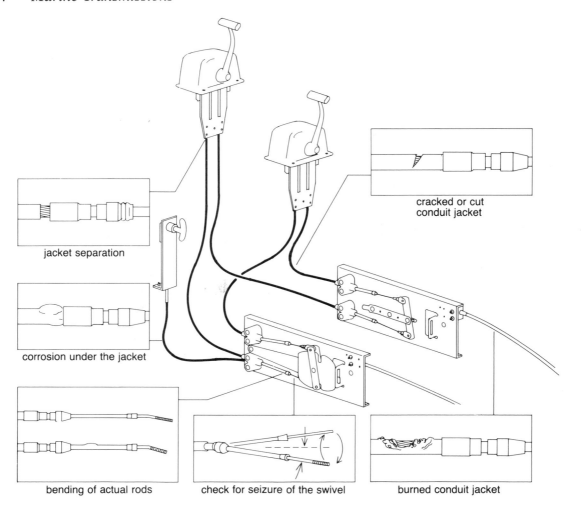

jacket separation

cracked or cut
conduit jacket

corrosion under the jacket

bending of actual rods

check for seizure of the swivel

burned conduit jacket

Figure 8-12. *Checking transmission control cables. (Courtesy Morse Controls)*

with lengthwise ridges all the way around) or a keyed shaft. In the latter case, do not lose the key down in the bilges when removing the coupling. The key will most likely stick in the shaft. If there is no risk of its falling out and getting lost, leave it there; otherwise hold a screwdriver against one end and tap gently until the end can be pried up and the key removed.

Some couplings are a friction fit on their shafts and should be removed with a proper puller (see Figure 8-14). This is nothing more than a flat metal bar bolted to the coupling and tapped to take a bolt in its center. The bolt screws down against the transmission output shaft, forcing off the coupling.

Transmission oil seals are a press fit into the rear transmission housing. Most seals consist of a rubber-coated steel case with a flat face on the rear end and a rubber lip on the front end (the end inside the transmission). A spring inside the seal holds this lip against the coupling face to be sealed.

Removing a seal from its housing is not always easy. If at all possible, the housing should be unbolted from the transmission and taken to a convenient workbench. This is often fairly simple on older boxes and boxes with reduction gears, but may not be feasible on many modern hydraulic boxes. The seal may be dug out with chisels, screwdrivers, steel hooks, or any

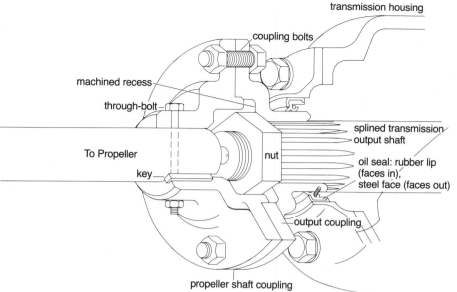

Figure 8-13. *Transmission oil seal and output coupling arrangement. The recesses machined into the face of the two coupling halves assist shaft alignment.*

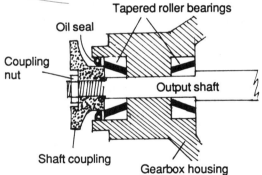

Figure 8-15. *(Top) A typical thrust-bearing arrangement and location of engine gearbox oil seal. (Bottom) Pre-loaded thrust bearings. (Courtesy Borg Warner)*

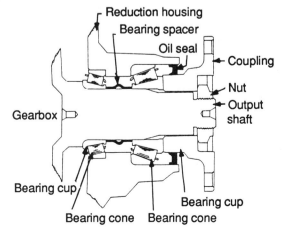

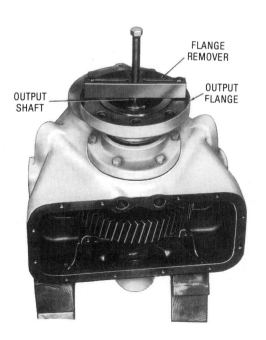

Figure 8-14. *Removing the output shaft coupling with a coupling puller. (Courtesy Allison Transmission)*

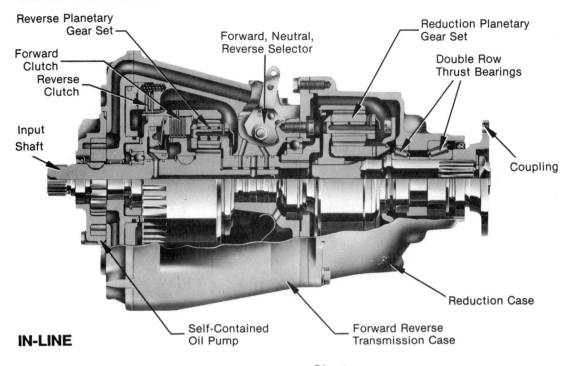

IN-LINE

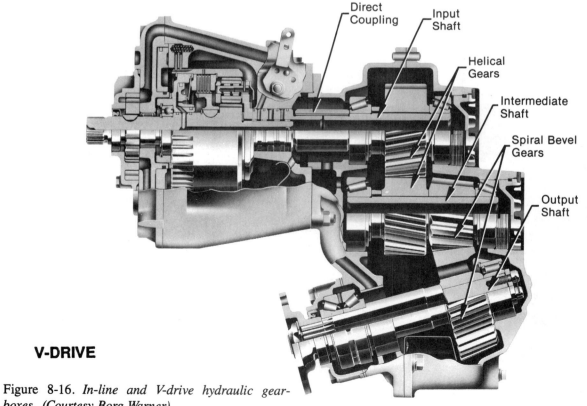

V-DRIVE

Figure 8-16. *In-line and V-drive hydraulic gear-boxes. (Courtesy Borg Warner)*

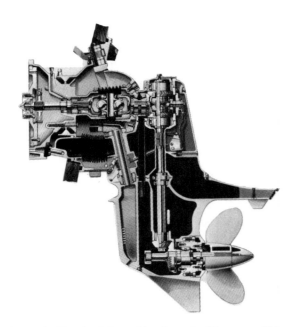

Figure 8-17. *An inboard/outboard. (Courtesy Volvo Penta)*

on which the coupling is mounted, turns in two sets of tapered roller bearings-one facing in each direction (see Figure 8-15). Between the two is a steel sleeve. When the coupling nut is pulled up, this sleeve is compressed, maintaining tension on the bearings and eliminating any play. Any time you undo the coupling nut, use a torque wrench and note the pressure that is needed to break loose the nut. When you reinstall the nut, tighten it to the same torque *plus 2 to 5 pounds*. In any event, the torque should be at least 160 pound-feet on Borg-Warner boxes, but the couplings should still turn freely by hand in neutral with only minimal drag. If the transmission needs a new spacer between the thrust bearings, a special jig and procedure are called for, and the whole transmission reduction gear will have to go to a professional.

On an older transmission in which the coupling nut is done up less tightly and restrained with a cotter pin, it is essential that the nut be properly replaced. The best approach is to moderately tighten the nut, make sure everything is properly seated, then back it off an eighth of a turn or so before inserting the cotter pin. The transmission should be put in neutral and the coupling turned by hand to make sure there is no binding.

other implement that comes to hand; it doesn't matter if the seal gets chewed up, as *long as the housing and shaft (if still in place) are unscratched.*

New seals go into the housing with the rubber lip facing into the gearbox, and the flat face outside. You must place the seal squarely into the housing then tap it in evenly using a block of wood and a hammer. If a seal is forced in cockeyed it will be damaged. The block of wood is necessary to maintain an even pressure over the whole seal face—*hitting a seal directly will distort it*. Push in the seal until its rear end is flush with the face of the transmission housing. Once in place, some seals will require greasing (there will be a grease fitting on the back of the gearbox), but most need no further attention.

Reassembly of a coupling and propeller shaft is the reversal of disassembly. Check alignment of the propeller shaft (see page 184) any time the coupling halves are broken loose and reassembled.

Many transmissions have what are called *preloaded* thrust bearings. The transmission output shaft,

V-Drives

A V-drive is simply an arrangement of gears that allows the engine to be installed "backward," placing it directly over the output shaft (see Figure 8-16). This enables far more compact engine installations to be made, and in particular enables sportfishing boats and other planing-type hulls to have the engine installed right in the stern of the boat, which is the best spot from the point of view of weight distribution.

Inboard/Outboards

As more and more diesels find their way into sportfishing boats, inboard/outboard transmissions are beginning to get coupled to them (see Figure 8-17).

Inboard/outboards are exactly what their name implies—an inboard engine coupled to an outboard-motor-type drive assembly and propeller arrangement. These units have definite advantages in planing craft, notably:

- Inboard/outboards allow an engine to be mounted in the stern of the boat, which is the best place in terms of weight distribution on many planing hulls.
- The outboard unit can be hydraulically pivoted up and down. This enables these boats to take full advantage of their shallow draft to run up onto beaches. It also provides infinite propeller depth adjustment for changes in boat trim, and makes trailering easy.
- The whole outboard unit turns for steering, greatly increasing maneuverability and removing the need for a separate rudder.
- There is no propeller shaft, stern tube, or stuffing box to leak into the boat.

Naturally, there are drawbacks. The extra gearing and sharp changes in drive angle absorb a little more power than a conventional transmission; the U-drives in the transmission tend to have a relatively high failure rate if driven hard; and above all, most outboard units are built of relatively corrosion-susceptible materials. If the boat is kept dry-docked between use, or the outboard unit raised out of the water, this latter point is not a great problem, but if the unit is left in the water it can be.

Observing all maintenance schedules is important, particularly oil changes, greasing of U-joints, and replacing those all-important zincs.

Chapter 9

Engine Selection and Installation

This chapter may seem of little relevance to boatowners who already have an engine installed in their boat. However, quite commonly problems with existing installations arise from either a poor matching of the engine to the boat's needs or from poor installation practices. In such a situation, these pages may throw some light on a longstanding problem.

SECTION ONE: ENGINE SELECTION

Matching the engine to its load and to its use are the primary considerations when installing a diesel engine in a boat. These in turn come down, to a large extent, to a consideration of how much horsepower you need and at what power-to-weight ratio?

Matching an Engine to Its Load

Diesels are susceptible to damage from overloading and underloading. When overloaded, generalized or localized overheating can lead to engine damage, up to and including seizure. The damage from under-

loading is in some ways more pernicious. It can arise from running at higher speeds with a low load (generally as a result of a mismatched propeller), or more likely, from repeated low-speed operation with little load. The latter is particularly common on auxiliary sailboats when charging a battery or running a refrigerator at anchor—it is not unusual to find a 50-h.p. motor carrying a $1/2$- to 1-h.p. load.

An underloaded engine takes time to reach proper operating temperatures, and at low speeds, it also tends to run unevenly due to the difficulties of accurately metering the minute quantities of fuel needed at each injection stroke. These two factors encourage the formation of sulfuric acid in the lubricating oil (see page 34), and carbon deposits throughout the engine. The cylinder walls are likely to become glazed, and piston rings will get gummed in their grooves, resulting in blow-by and a loss of compression. Valves may stick in their guides, while carbon will plug up the exhaust system. A carbon sludge will form in the oil, and if you neglect oil change procedures, the sludge will eventually plug sensitive oil passages and lead to bearing failure.

Repeated running of a diesel engine at low loads is a destructive practice, which greatly increases maintenance costs and reduces engine life.

PLANING SPEED CHART CONSTANTS

C	Type of Boat
150	average runabouts, cruisers, passenger vessels
190	high-speed runabouts, very light high-speed cruisers
210	race boat types
220	three-point hydroplanes, stepped hydroplanes
230	racing power catamarans and sea sleds

Figure 9-1. *From* Propeller Handbook, *by Dave Gerr, International Marine Publishing.* *(Courtesy International Marine Publishing Co.)*

How Much Horsepower Do You Need?

Does your boat have a displacement or a planing-type hull? A displacement hull is one that remains immersed at all times, whereas a planing hull develops hydrodynamic forces at speed that enable it to move up onto the surface of the water.

A displacement hull has a pre-determined top speed (defined as *hull speed)* more or less irrespective of available power. This top speed is governed by certain physical properties of the waves the boat makes as it passes through the water (see the sidebar) and is approximately 1.34 times the square root of the boat's waterline length. A clean-hulled displacement craft can be driven at close to its hull speed in smooth water by a relatively small engine, but as hull speed is approached, drag (resistance) increases rapidly and any additional speed can only be gained by a disproportionate increase in power (therefore fuel burned).

A planing hull, on the other hand, breaks free of the constraints imposed by the waves it generates. A certain minimum amount of power is required to come up to a plane. Thereafter, the boat's top speed is at least in part related to available power (see Figures 9-1 and 9-2).

Various formulas have been derived for determining the horsepower requirements of displacement and planing hulls. Two excellent sources are *Skene's Elements of Yacht Design*, by Francis S. Kinney, published by Dodd, Mead & Co. (see the sidebar), and the *Propeller Handbook*, by Dave Gerr, published by

International Marine. The formulas will enable you to take into account the effects on boat speed of a foul bottom, wave action, head winds and other factors.

Only in exceptional circumstances will you find that a displacement hull requires more than 1 h.p. per 500 pounds (fully-loaded) displacement (see Figure 9-3). We have a 30-h.p. engine in a 30,000-pound boat and have always found it adequate, but then we have a very efficient variable-pitch propeller. We are operating at the lower end of the power requirement for our boat.

Wave Theory

As a boat moves through the water, it makes waves, which behave according to certain physical laws—the faster the waves move, the wider apart they are spaced. The distance from one wave crest to the next is the *wavelength*. As a boat picks up speed, so too does its associated wave formation, and the faster it goes, the farther apart the waves become. A table can be constructed showing the speed of waves of any particular wavelength, or the wavelengths of waves of any particular speed (see Figure 9-4).

A boat eventually reaches a speed at which the length of its associated wave formation is the same as its waterline length—one wave crest is at the bow (the bow wave) and the next is at the stern (the stern wave). If the boat were to go any faster, its wave formation would also speed up, therefore

PLANING SPEED

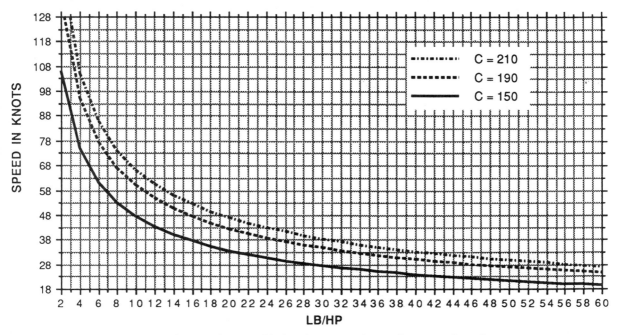

Figure 9-2. *This chart shows the speed attainable by planing craft as a function of available shaft horsepower. See Figure 9-1 to estimate the appropriate C value with which to enter the table. From* Propeller Handbook. *(Courtesy International Marine Publishing)*

lengthen; the boat would move ahead of its stern wave, its stern would sink into the trough between the bow and stern waves, and its bow would appear to be climbing the bow wave. In a sense, the boat would be dragging its stern wave behind it, requiring tremendous amounts of power. As a consequence, the maximum speed of a displacement hull (its *hull speed*) is determined by the speed of the wave formation, with a wavelength equal to the waterline length of the boat. The longer a boat's waterline length, the farther apart its bow and stern waves will be, therefore the faster it can go. This is why a sailboat with long overhangs can move faster when it is heeled—the heeling increases its waterline length.

Figure 9-4 shows how dramatically wavelengths increase with small increases in speed. To double a displacement boat's hull speed from 5 to 10 knots requires a fourfold increase in the waterline length.

Only in the most exceptional circumstances, such as when a boat surfs down the face of a wave, can a displacement hull exceed its hull speed. The more closely hull speed is approached, the greater the increase in power required for a given increase in speed. At around 75% of hull speed, the boat is extremely efficient, but beyond this point the additional fuel burned becomes increasingly disproportionate to any increase in speed, due to the rapid rise in drag.

A planing hull, on the other hand, breaks free of its own wave formation by moving up onto the surface of the water (see Figure 9-5). The moment at which this occurs is often felt as a sudden surge in speed as the boat accelerates away from its stern wave, barely skimming the surface of the water.

DISPLACEMENT SPEED—INCLUDING SEMIDISPLACEMENT

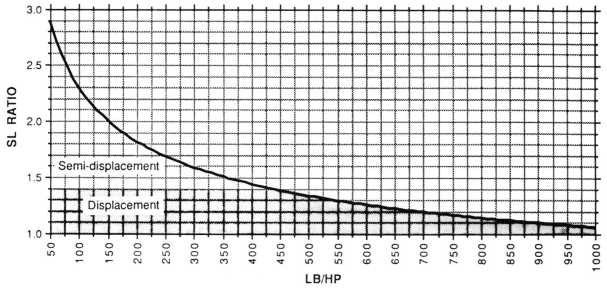

Figure 9-3. *This chart shows the power necessary to achieve a boat's known maximum speed-length ratio. It would be tempting to conclude from the chart that even a heavy-displacement hull can achieve SL ratios of 1.5 or higher, given enough power, but in practice, such an attempt would be unfeasible. For most moderate-to-heavy-displacement vessels, incorporating more than one horsepower per 500 pounds or so of displacement in an effort to achieve SL ratios higher than 1.3 to 1.4 is neither practical nor economical. Heavy hulls designed with planing or semiplaning underbodies may be driven to semidisplacement speeds, but only at a great cost in fuel consumption and power. From* Propeller Handbook. *(Courtesy International Marine Publishing)*

VELOCITY IN KNOTS	WAVE LENGTH IN FEET
1	0.56
2	2.23
3	5.01
4	8.90
5	13.90
6	20.0
7	27.2
8	35.6
9	45.0
10	55.6
11	67.3
12	80.1
13	94.0
14	109.0
15	125.2

Figure 9-4. *Table of periods and lengths of sea waves.*

BHP, SHP, and Auxiliary Equipment

The figures given for horsepower in most engine manufacturers' specifications are measured before adding the transmission, any reduction gears, and the propeller shafting. This is the engine's Brake Horse Power (BHP). But the figures we have derived so far are those needed *at the propeller*, otherwise known as Shaft Horse Power (SHP). However, since powertrain losses are generally only 3 to 5 percent of the BHP, except in special circumstances, the difference between BHP and SHP can be largely ignored.

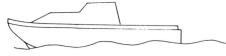

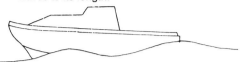

A 32-foot (waterline length) boat moving at 4 knots. There will be approximately 3½ waves to its length.

The same boat moving at 7½ knots. There will be approximately one wave to its length. This boat is moving at hull speed for displacement boats of that length.

Figure 9-5. *Boat speed and wave formations.*

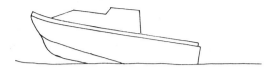

The boat is now moving at, say, 12 knots. It has moved up onto the surface of the water and ahead of its own wave formation.

Determining the Horsepower Requirements of an Auxiliary Sailboat Using the Formulas in *Skene's Elements of Yacht Design*

Start from the boat's waterline length—let us assume 32 feet—and a speed/length ratio of 1.34. This gives a hull speed of:

1.34 × the square root of LWL = 1.34 × the square root of 32 = 7.58 knots for a waterline length of 32 feet.

The graph shown in Figure 9-6 is entered on the bottom line at 1.34 and traced upward to the lower curve for light-displacement hulls, and to the upper curve for heavy-displacement hulls. Let us assume a heavy cruising boat of 26,000 pounds (note that this weight should include all stores normally on board). Using the upper curve, we move horizontally to find the resistance in pounds for each long ton of displacement (a long ton = 2,240 lbs.). For a speed/length ratio of 1.34, the resistance is 55 pounds per long ton. 26,000 lbs. = 11.6 long tons, therefore total resistance at hull speed (7.58 knots) for this hull is:

$$11.6 \times 55 = 638 \text{ lbs}$$

Effective horsepower (EHP) is given by the formula:

EHP = resistance × speed × 0.003
= 638 × 7.58 × 0.003
= 14.5 h.p.

Propellers are notoriously inefficient at transmitting power. Kinney uses the following factors:

Folding two-blade: 10%
Auxiliary two-blade: 35%-45%
Fixed three blade: 50%

Let us assume an average auxiliary two-bladed propeller with a 40% rating. We arrive at the following horsepower to drive our boat at hull speed:

$$14.5/0.40 = 36.25 \text{ h.p.}$$

Kinney also adds 33% for adverse conditions, to give a maximum power requirement, in this example, of 48 h.p. (this is 542 pounds per horsepower, which is pretty much the same as the figure derived from Dave Gerr's chart using a speed/length ratio of 1.34). Note that if we had used a speed length ratio of 1.00 with this boat, giving a top speed of 5.7 knots, the horsepower requirement would have only worked out to be 9.5. Put another way, this hull can be pushed at 5.7 knots by 9.5 h.p., but will need 48 h.p. to move at 7.58 knots. This dramatically illustrates the increase in drag, therefore fuel consumption, as hull speed is approached.

Figure 9-6. *From Francis S. Kinney's* Skene's Elements of Yacht Design. *(Courtesy Dodd, Mead & Co.)*

The effect of belt-driven auxiliary equipment is often of more concern. The DC loads on boats are steadily increasing from year to year as boatowners add more and more gadgets. In order to keep up with this burgeoning load, the tendency is to fit more and more powerful alternators—130-amp and 160-amp models are becoming quite common. At full load, these absorb more than 2 h.p. from an engine. Engine-driven refrigeration compressors can make similar demands. On large engines, the impact of such devices is negligible, but on an engine fitted to a small auxiliary sailboat, it may be considerable and will certainly need to be taken into account. Every horsepower absorbed by auxiliary devices is lost at the propeller shaft. The engine size may need to be increased to compensate for these losses.

When comparing manufacturers' specifications, you must differentiate between an engine's horsepower rating in *continuous duty*, and in *intermittent duty*. An intermittently rated engine is designed to be operated at full power for limited periods only. Auxiliary sailboats, which rarely use their engines at full power for prolonged periods, can use an intermittent-duty rating for choosing a suitably-sized engine. Many other boats, however (e.g., ocean-going motorsailers or sportfishing boats), will need to have an engine based on its continuous-duty rating.

Power-to-Weight Ratio

Having determined the boat's power requirement, we now need to match the engine to the boat's use. Let us consider two extreme examples.

A Heavy-Displacement Cruising Sailboat.
Engine horsepower can be determined using Dave Gerr's graph or Kinney's formulas. Weight is not critical. The owner is principally interested in reliability, simplicity, longevity, and ease of maintenance. This suggests a relatively slow-turning naturally-aspirated 4-cycle marine diesel. The slow speed promotes a long life, whereas natural aspiration does away with all the complications of a turbocharger and aftercooler. Every part of the engine will be easily accessible, and even removing the cylinder head will not involve too much work.

This owner might add a few more requirements. Since plans involve long-range cruising, the engine should have a hand-start capability in case the battery goes dead in some remote anchorage. For similar reasons, a cold-start device based on adding oil to the cylinders is preferred over one (such as glow-plugs) that requires battery power (a difficult requirement to meet these days).

This particular owner expects to encounter severe weather, so he doesn't want access hatches in the cockpit that might leak. In effect, once the engine is installed, the cockpit will be built over it—the only way to get the engine out again will be in pieces. Therefore it must be possible to *completely* rebuild the engine *in situ*. Such a requirement calls for an engine with replaceable wet-type cylinder liners, and if possible, access to the connecting rods and caps through hatches in the crankcase, instead of having to drop the pan (see Figure 9-7).

A Lightweight Sportfishing Boat.
The maximum amount of power for the minimum weight is critical. Tests on sportfishing boats have shown that outboard motors will give higher top speeds and greater fuel economy than more powerful inboard gasoline engines. (See, for example, *Boating* magazine, February 1986, page 102, "Outboard v. Inboard." I am not aware of any similar tests with diesel engines.) The critical factors at work here are the extra weight of the inboard engine(s), and its placement farther forward than the transom-mounted outboards.

This gives some idea of the supreme importance of maximizing the power-to-weight ratio in this kind of craft. This also indicates the use of relatively high-revving diesels. All other things being equal, if one engine the same size as another is run twice as fast, it can suck in twice as much air and develop twice as much power. In order to further boost power, a turbocharger and aftercooler will be needed (see Figure 9-8). A V-drive would almost certainly help to keep the engine's weight in the most desirable location.

The owner of this boat expects to dock at a marina every night. There is no need for hand-starting, decompression levers, or oil-based cold starting devices—the far more convenient glow plugs will be used. The more complicated maintenance procedures are likewise no problem because the marina's skilled mechanic will carry out the maintenance. The engine(s), housed under a console on the main deck, can be easily removed for major overhauls.

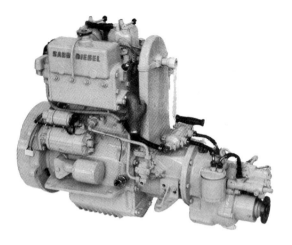

Figure 9-7. *A traditional marine diesel—Sabb 2JZ.* *(Courtesy Sabb Motor A.S.)*

Figure 9-8. *A turbocharged Caterpillar 3208 diesel.* *(Courtesy Caterpillar Tractor Co.)*

SECTION TWO:
THE POWER TRAIN

The power train is a general expression to describe transmission, reduction gear, and propeller arrangements. Some mention has already been made of the power losses created by reverse and reduction gearing. You might wonder why manufacturers do not use a direct drive and dispense with the power losses of a reduction gear. This is, in fact, done on some large ships where the engines turn over at speeds as low as 300 rpm. In fact, some of these engines have two camshafts and two sets of valves so that they can be stopped and restarted backwards for reverse, thus eliminating the need for a reverse gear.

High-revving diesel engines in lightweight, high-speed boats also frequently have a direct drive to the propeller, though a reverse gear is required. But on slower, heavier boats, especially displacement hulls, the high propeller speeds of a direct drive would create excessive propeller *slip* and *cavitation*.

Slip

A propeller is specified according to its diameter, its *pitch*, and whether it is left-handed or right-handed. The latter merely indicates which way the propeller turns to produce forward thrust; the diameter is self-evident. Pitch is a little more complex.

The blades of a propeller are set at an angle to its hub (*boss*, or center). As the propeller turns, the angled faces cut into the water, and in the process create pressure on the rear side of the blade and a vacuum on the front edge (just as with the upper edge of an airplane wing). The two forces together impart motion to the boat.

In a perfect environment, no motion would be imparted to the water, and the propeller would move ahead by the total amount of the deflection of its blades. In practice, of course, some movement forward of the propeller (and therefore the boat) takes place, at the same time that water is driven back past the propeller. In the perfect environment, the greater the angle of deflection of the propeller blade (within reason), the farther the propeller would move in one revolution. This theoretical distance, measured in inches, is the propeller's *pitch*. A 12-pitch propeller would move forward 12 inches, a 16-pitch propeller, 16 inches. In real life, the 12-pitch propeller might move the boat forward 6 inches per revolution. The difference between this figure and the theoretical movement represents the degree of *slip*—in this case 50%.

The degree of slip obviously varies with circumstances but is generally between 20% and 50%. When the boat is dead in the water, it needs a considerable push to overcome its inertia and get it moving, and this will show up as a high degree of slip. Once the boat gains momentum, slip lessens. Slip is much greater (approaching 100% in extreme conditions) when the boat is punching into a head sea or a strong wind than in smooth water.

Cavitation

Cavitation occurs when the vacuum formed on the front face of a propeller blade becomes excessive and causes bubbles to form. These can cause pitting of the propeller blade and vibration. Cavitation occurs when a propeller turns too fast in a specific situation. It is often confused (including in the first edition of this book) with *ventilation*, which arises when the propeller sucks air down from the surface of the water, which is likely to be the result of running the propeller at too high a speed, or of having it insufficiently submerged.

Propeller Selection

Reduction gears alter the engine's maximum shaft speed to produce a propeller speed that results in minimal slip and cavitation for a particular application. The slower the shaft speed, the greater must be the diameter and/or pitch of a propeller to maintain thrust. In general, large, slow-turning 3-bladed propellers are more efficient when the boat is *under power* than small high-revving 2-bladed propellers, but the former generate far more resistance when the boat is *under sail*. The upper limit on propeller diameter is generally determined by the space available beneath a hull—the blade tips need to clear the hull by at least 15% of the overall diameter of the propeller (see Figure 9-9).

The optimum balance between propeller diameter, pitch, type, and shaft speed is clearly a difficult one to achieve. It can only be determined by reference to specific boat use, since two identical boats may have different requirements. For example, one sailboat may be used for racing around the buoys, whereas another is used for cruising in an area that has light and variable winds. The former owner will want a propeller that gives the least resistance when not in use and will likely choose a high-revving, small-diameter, two-bladed folding propeller and accept the inevitable inefficiencies under power. The latter owner expects to do quite a bit of motorsailing and will be looking for efficiency under power—a fairly large 3-bladed propeller will likely be chosen (with an appropriate reduction gear), despite the increase in drag under sail.

If a propeller is too powerful for the engine turning it (too large a diameter, excessive pitch, or both), the engine will be overloaded. It will be unable to attain full speed and most likely will emit black smoke and potentially carbon up its valves. On the other hand, if the propeller is undersized (too small a diameter, inadequate pitch, or both), it will not produce sufficient thrust, and the engine is likely to overspeed while the boat fails to attain its designed speed.

Unfortunately, it is common practice to fit the most powerful propeller that an engine will handle in optimum conditions—smooth water, a lightly-laden boat, and a clean drag-free hull. This exaggerates the performance of the boat under power but overloads the engine under normal operating conditions. *The chosen propeller must allow the engine to reach at least 90% of its rated speed under full load, taking into account adverse conditions, belt-driven auxiliary equipment, and all stores likely to be on board.*

In the final analysis, propeller selection is partly science, partly art (or experience). Once you have

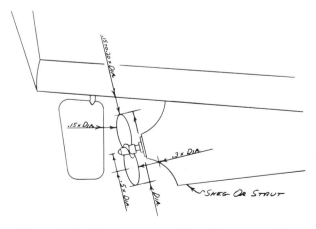

Figure 9-9. *Minimum propeller clearances. From* Propeller Handbook. *(Courtesy International Marine Publishing)*

determined your available SHP and decided the general characteristics of the propeller you want, you will need to discuss the possibilities with an experienced propeller dealer or naval architect. Even then, you may have to experiment and change the propeller more than once.

Feathering Propellers. Perhaps the best of all possible worlds for sailboats is one of the new breed of feathering propellers, such as those made by Max-Prop (see Figure 9-10). Each blade has on its base a bevel gear, which engages with a central beveled cone gear mounted on the propeller shaft. The whole unit is enclosed in a case (the spinner).

When the engine is first cranked, the blades will be in a feathered position. If the forward gear is engaged, the propeller shaft turns while the case and blades lag due to inertia. The initial torque of the turning shaft and cone gear rotates the propeller blades in the case via the bevel gears. The blades have a preset stop that prevents them from turning beyond a certain point. This determines their pitch. Once the stop is reached, the propeller shaft spins the whole unit, including the case.

When the engine is shut down, water pressure on the blades forces them back to the feathered position. In reverse, the propeller shaft once again drives the blades to their full pitch, but in the other direction, before spinning the whole unit. This ensures maximum efficiency in reverse.

This is a very neat design indeed, but expensive. The pitch can be fine-tuned for any boat and conditions by altering the position of the pitch stop, but to do this, the boat needs to be hauled and the case disassembled.

(Note: Quite the best book on propeller installations is Dave Gerr's *Propeller Handbook* already mentioned.)

SECTION THREE: MOUNTING THE PROPELLER SHAFT

The propeller shaft runs in some kind of a stern tube extending through the hull. It contains at least one bearing. There may or may not be an externally mounted strut, depending on how far the unsupported section of the shaft protrudes from the stern tube (see Figure 9-11). The inner end of the stern tube has a seal to keep water out of the boat—generally a *stuffing box*, or *packing gland*.

Struts and Cutless Bearings

Any strut must be well-fastened to the hull with a generous backing block to spread the loads. Looseness or flexing of the hull around the strut mounting will lead to vibration. The vibration will damage stern-tube and transmission bearings, leading to failure of the stern-tube shaft seal.

The strut, or the stern-tube if no strut is fitted, will incorporate a cutless-type bearing—a ribbed rubber sleeve that supports the propeller shaft (Cutless is a registered trademark of L. Q. Moffitt, but has come to be used as a general name for this type of bearing). The shaft needs to be a close, but not a tight, fit in this sleeve. If you utilize an existing bearing, flex the propeller shaft up and down, and from side to side. If there is more than minimal movement, the bearing should be replaced. If the cutless bearing is worn on only one side, this is a sure sign of previous engine misalignment.

You must remove the shaft to renew a cutless bearing. Most bearings are a simple sliding fit, locked in place with set screws. Once you have loosened the set screws, knock out a strut-mounted bearing from the inner side of the strut. To do the same on a hull-mounted bearing (e.g., one set into the deadwood), you have to remove the stuffing box from the inner end of the stern tube.

The bearing quite commonly refuses to budge. There is a limit to how hard you can beat on it. I wrap multiple layers of tape around a hacksaw blade to form a handle then cut two longitudinal slits in the bearing and pry out a section, which allows the rest to be flexed inwards and removed. (Perhaps a reader has a better suggestion: This is time-consuming, especially on metal-shelled bearings, and carries the risk of accidentally cutting into the surrounding strut or stern tube.)

New cutless bearings come in naval brass, stainless steel, or fiberglass sleeves. The latter work just as well as the metals and are not susceptible to corro-

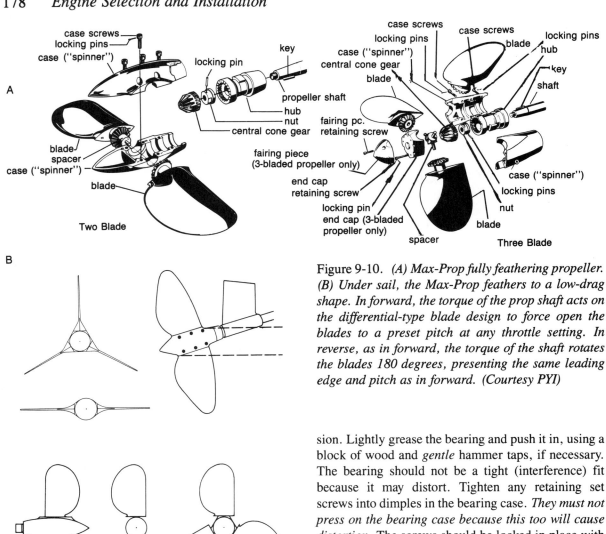

Two Blade

case screws
locking pins
case ("spinner")

locking pin

key

propeller shaft

hub
nut
central cone gear

blade
spacer
case ("spinner")

blade

case screws
locking pins
case ("spinner")
central cone gear
blade

case screws
blade

locking pins
hub

key
shaft

fairing pc.
retaining screw

fairing piece
(3-bladed propeller only)

end cap
retaining screw

locking pin
end cap (3-bladed
propeller only)

spacer

case ("spinner")
locking pins
nut
blade

Three Blade

Figure 9-10. *(A) Max-Prop fully feathering propeller. (B) Under sail, the Max-Prop feathers to a low-drag shape. In forward, the torque of the prop shaft acts on the differential-type blade design to force open the blades to a preset pitch at any throttle setting. In reverse, as in forward, the torque of the shaft rotates the blades 180 degrees, presenting the same leading edge and pitch as in forward. (Courtesy PYI)*

sion. Lightly grease the bearing and push it in, using a block of wood and *gentle* hammer taps, if necessary. The bearing should not be a tight (interference) fit because it may distort. Tighten any retaining set screws into dimples in the bearing case. *They must not press on the bearing case because this too will cause distortion.* The screws should be locked in place with Loctite, or something similar.

A cutless bearing will last for years if the engine is properly aligned (see below) and the bearing is adequately lubricated. Water flowing up the grooves in the rubber lubricates the bearing. When a bearing is mounted in an external strut, there is plenty of water flow, but when it's mounted in the deadwood of an auxiliary sailboat, an extra lubrication channel into the bearing is often needed.

Keeping the Water Out

The inner end of a stern tube needs to form a more-or-less watertight seal around the propeller shaft. The

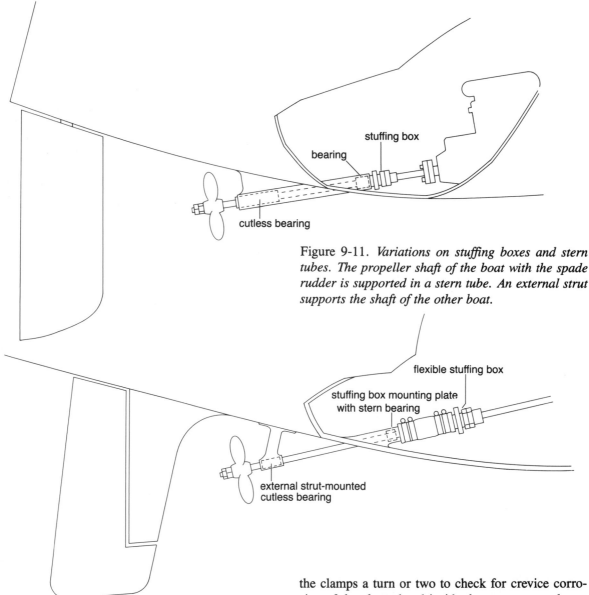

Figure 9-11. *Variations on stuffing boxes and stern tubes. The propeller shaft of the boat with the spade rudder is supported in a stern tube. An external strut supports the shaft of the other boat.*

traditional way of doing this is to fit a *stuffing box* (*packing gland*), which can either be rigidly mounted—threaded onto the end of the stern tube—or flexibly mounted—fastened to the end of the stern tube with a length of hose (see Figure 9-12). In the latter case, be sure to use two all-stainless steel hose clamps at both ends of the hose and to check the hose and clamps carefully every year. In particular, undo the clamps a turn or two to check for crevice corrosion of the clamp band inside the worm screw housing. A failure of the hose on a flexible stuffing box can let in large amounts of water *fast*.

A stuffing box consists of a small cylinder fitted around the propeller shaft, forming a close fit at its lower end. A large *packing nut* or *clamp plate* makes another close fit around the shaft and closes off the top of the cylinder. Rings of greased flax are pushed down into the cylinder around the shaft (generally three or four rings). This is the packing. A metal sleeve—the *compression spacer*—is placed on top (or

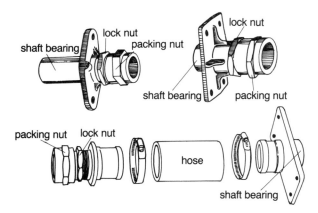

Figure 9-12. *Rigid stuffing boxes (top) and a flexible stuffing box (bottom). (Courtesy Wilcox Crittenden)*

incorporated into the underside of the clamp plate). Tightening the nut or clamp plate compresses the packing, squeezing it out against the sides of the cylinder and up against the shaft, sealing the shaft (see Figure 9-13).

Packing comes as a square-sided rope in a variety of sizes: 3/16" (4mm), 1/4" (6mm), 3/8" (8mm), etc. It is important to match the packing to the gap between the shaft and cylinder wall. You can buy packing as preformed rings to match the stuffing box (the best option) or by the roll. When cutting rings off a roll, make about five tight wraps around the propeller shaft at some convenient point, then cut diagonally across the wraps with a very sharp knife.

Packing Adjustment and Replacement

At this point, it is convenient to get a little ahead of myself and assume the engine is installed and running. A stuffing box is meant to leak. When the shaft turns, it needs two or three drops a minute to keep lubricated. If the leak is worse than this, tighten the nut or clamp plate to compress the packing a little more. If a greaser is fitted, pump in a little grease first. Tighten down a nut no more than one-quarter turn at a time; tighten evenly the two nuts of a clamp plate.

Let the engine run for a couple of minutes with the transmission in gear, then shut it down and immediately feel the stuffing box and adjacent shaft. If they are hot, the packing is too tight. A little warmth is acceptable for a short while as the packing beds in, but any real heat is completely unacceptable. It is quite possible (and common) to score grooves in shafts by overtightening the packing, in which case the shaft will never seal and will have to be replaced (sometimes it can be turned end for end to place a different section in the stuffing box).

If the shaft cannot be sealed without heating, the packing needs replacing. It should, in any case, be renewed every year, since old packing hardens and will score the shaft when you tighten the packing. The hardest part of the job is generally getting out the old packing—you have to remove all traces of this or the new packing will never seat properly. With a deep, awkwardly-placed stuffing box it can be almost impossible to pick out the inner wraps of packing with screwdrivers and ice picks. A special tool is needed, consisting of a corkscrew on a flexible shaft (see Figure 9-14). If this tool is not available, do not start digging into the packing, especially if the boat is in the water. Appreciable quantities of seawater may start to come in as you remove the packing, in which case speed is of the essence.

When fitting new rings of packing, lubricate each with a Teflon-based waterproof grease before installation, and tamp it down before adding the next ring. I use some short pieces of pipe slit lengthwise, slipped around the shaft, and pulled down with the packing nut or clamp plate to *gently* pinch up the inner wraps (not too tight or the shaft will burn). Stagger the joints from one wrap to another by about 120 degrees.

Graphite Packing Tape

Graphite packing tape can be used in place of flax. It comes as a fragile reel of tape that is wrapped around and around the shaft until a sufficient thickness is built up to fill the space between the shaft and stuffing box cylinder. Slide these wraps into the stuffing box and repeat the process until the box is full. Compress the packing by tightening the nut or clamp plate then add more packing. The total uncompressed width of the rings of tape should be about one-and-a-half times the depth of the stuffing box.

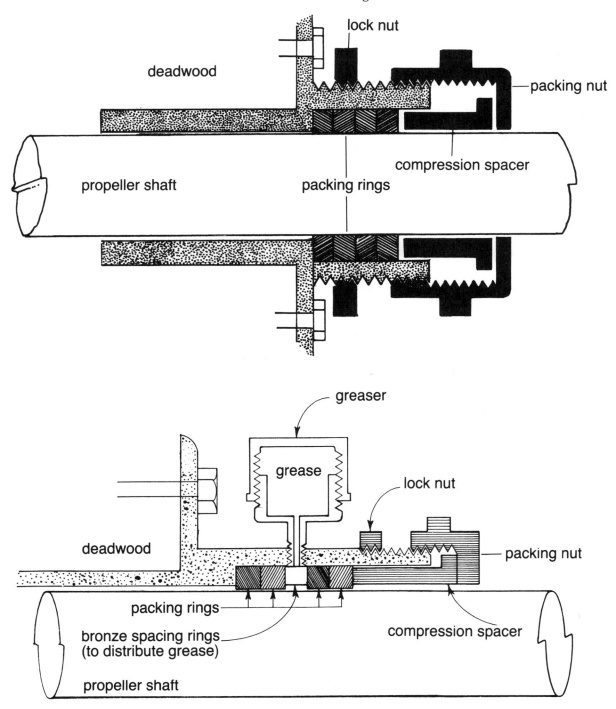

Figure 9-13. *Cross section of a rigid stuffing box (top). Cross section of a stuffing box equipped with a grease fitting (bottom). You can replace this standard greaser with a grease nipple or remote greaser.*

Figure 9-14. *Removing the packing from a deep, awkwardly placed stuffing box should be reserved for this special tool. Screwdrivers and other substitutes may not remove the last few turns of packing.*

Graphite packing tape is far superior to flax in many ways. It crushes to form an excellent seal. Since graphite is a lubricant, there is less risk of burning the shaft. If extra packing is needed, it is simply added to the box—you do not have to clean out the stuffing box. However, it has two drawbacks: The tape is quite expensive and sometimes hard to find, and graphite is high on the galvanic table and thus may promote corrosion. For this reason, its use is generally discouraged, but we have had it in our stuffing box for eight years without problems.

Rotary Seals

Rotary seals are new on the boating scene and are used in place of a stuffing box. A rubber boot, in which is embedded a solid stationary seat, is clamped to the stern tube. A second rubber boot with a hard ring molded into it (the rotating seal) is slid up the propeller shaft (the coupling must first be taken off) until the rotating seal mates with the stationary seat. This boot is pushed up a little more to maintain a gentle pressure between the seal and seat. The seal is then clamped to the propeller shaft (see Figure 9-15).

As the shaft turns, so does the rotating seal. The smooth faces of the seal and seat prevent any leaks. A small amount of water—the occasional drop—is necessary to lubricate the faces of the seal; without it the seal will heat up and self-destruct. Sometimes lubrication is inadequate, especially on wooden boats with a long stern tube through the deadwood. In this case, there is provision for water from the engine's raw-water circuit to be injected into the seal assembly. The

seal should be checked when the boat is under way to make sure that it is not heating up: the motion of a boat through the water can create a vacuum around the stern tube, drawing the water out of the stern tube.

After a haul-out, *it is essential to ensure that a rotary seal is properly lubricated before cranking the engine*. This is done by pulling back the rubber boot on the propeller shaft to release trapped air and holding it back until water spurts out. The only maintenance on a rotary seal is checking the hose clamps and rubber boots annually to ensure that they are not deteriorating. As with a flexible stuffing box, a failure can let in alarming amounts of water. Properly lubricated, the seals last almost indefinitely.

SECTION FOUR: ENGINE BEDS AND ALIGNMENT

Engine beds

By now you should have selected an engine, power train and propeller for your boat, and you should have the stern tube installation completed. It is time to get on with the engine installation. I find it useful to make a plywood template to match the engine feet, with extensions to represent the transmission and reduction gear, and with the position of the output coupling on the reduction gear accurately located. If space below or above the engine is critical, a cardboard pan or engine block can be added to indicate the maximum extension of the engine in any direction. This template can be maneuvered around far more easily within the confined spaces of the boat than the engine itself, enabling the precise positioning of the engine, therefore the engine beds, in relation to the stern tube to be worked out.

Engines are extremely rigid—necessarily so, because the crankcase and block are designed to withstand all the internal stresses and any tendency to flex generated by the moving parts. However, all boats flex to a degree. If the engine beds flex with the boat and the engine is rigidly bolted down, some serious problems can arise. The engine crankcase may be distorted (even if only by a thousandth of an inch or so), imposing severe stresses on the crankshaft. These stresses can cause the crankshaft to fail, and the

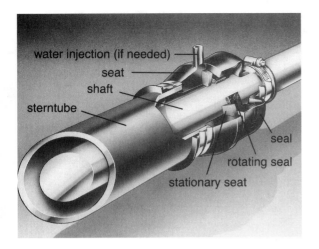

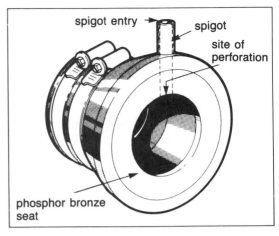

Figure 9-15. *The low maintenance of rotary seals make them an attractive alternative to conventional stuffing boxes. (Courtesy DEEP SEA SEALS and Halyard Marine)*

engine and propeller shaft alignment to be thrown out, leading to excessive vibration and wear in the bearings of the transmission and stern tube.

Engine beds must transfer the stresses generated by an engine to the hull of the boat, while at the same time they stiffen the hull as much as possible so that its deflection due to wave action is not transmitted to the engine. The longer the beds in general, the greater their effect, although clearly they must *in themselves* be solid enough to resist the bending forces applied to them. A thin piece of wood, for example, fastened to the hull will flex along with the hull and serve no purpose, regardless of length.

Think carefully about spreading engine-bed loads up the sides of the hull. *Rigid engine beds in an otherwise flexible hull can set up localized hard spots, which will result in cracked fiberglass laminates and wooden frames.* Fortunately, most engine beds are installed well down toward the keel, where the hull is already at its stiffest.

The engine bed must incorporate some form of a drip pan to catch oil and fuel-oil leaks from the engine. This is a U.S. Coast Guard regulation. Allowing engine oil to leak into the general sump and be discharged overboard is strictly illegal and carries heavy penalties. This regulation is enforced.

Couplings

Couplings should be keyed to their shafts then pinned or through-bolted so that they cannot slip off (see Figure 9-16). The practice of locking a coupling with a couple of set screws is not seaworthy—if the screws slip when the transmission is in reverse, the propeller and shaft are likely to pull out of the boat and let the ocean pour in through the open hole. If your coupling is held with set screws, at the very least you should drill and tap the screws into the shaft a thread or two and lock them with Loctite. It is a good idea to place a stainless steel hose clamp around all shafts just ahead of the stuffing box so that if the coupling ever works loose, the clamp will still stop the shaft from leaving the boat.

Accurate engine alignment (see below) depends on having a straight propeller shaft and on having the two coupling halves exactly centered on, and square to, their shafts (see Figure 9-17). A coupling should be fitted to its shaft and machined to a true fit in a lathe before the final shaft and coupling installation in the boat.

Engine alignment

In a rigid enough hull, it is still hard to beat the smooth running of a properly installed and aligned *rigidly* bolted engine and *rigid* propeller shaft coupling. However, it takes a good deal of experience to achieve satisfactory results, and the tendency today is overwhelmingly to use flexible feet and a flexible propeller coupling.

Accurate alignment is not only essential to the smooth operation of rigidly mounted engines, it is also important where flexible engine feet and couplings are used. The purpose of these flexible feet and couplings is to absorb wave-induced flexing in the hull, not to compensate for sloppy alignment. The alignment on all boats must be done with the boat in the water and loaded with all normal equipment and stores. After a haul-out, wooden boats should be left in the water a week before checking the alignment. This gives the hull a chance to settle to its normal shape.

To check the alignment on your engine, separate the propeller shaft coupling halves. If you have a long

Figure 9-16. *Shaft couplings may be keyed and pinned in place (top and center), or restrained by set screws seated in dimples in the propeller shaft (bottom).*

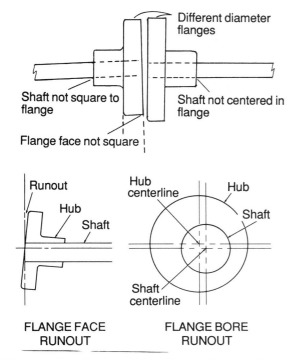

FLANGE FACE RUNOUT

FLANGE BORE RUNOUT

Figure 9-17. Shaft and coupling problems affecting engine alignment. (Courtesy Caterpillar Tractor Co.)

run of propeller shaft unsupported by a bearing, it will sag and some compensation is required for this. The correct procedure is to calculate half the weight of the protruding shaft, add to it the weight of the coupling, then pull up on the shaft by this amount with a spring scale of the type used to weigh fish (see Figure 9-18). In practice, you can generally flex up and down by hand the shaft of a smaller yacht to get a very good idea of the centerpoint, then support the shaft with an appropriately sized block of wood. Cut a notch into the wood to hold the shaft.

One half of the coupling should have a machined step that fits into a recess in the other half. Bring the two halves back together—the step should slip into the recess cleanly and without snagging at any point. If it does not, the shafts are seriously misaligned, and the engine needs jacking around until the halves come together. Now bring them almost into contact and use a feeler gauge to measure the gap between the coupling halves at the top, bottom, and both sides (see Figure 9-19). The difference from any one point to

another should not exceed 0.001″ per inch of coupling diameter (for example 0.003″ on a 3-inch diameter coupling).

Rotate the propeller shaft 180 degrees while holding the transmission coupling stationary and measure the clearances again. If the widest gap is still in the same place and the clearances are not within tolerance, the alignment needs adjusting. If the widest gap also has rotated 180 degrees, either the propeller shaft is bent or its coupling half is not squarely on the shaft—it should really be trued by a machine shop before proceeding any further.

Fine-tune the alignment by moving the engine around. Some engines have adjustable feet, which greatly simplify the procedure. Others require placing thin strips of precisely machined metal (shims) under the feet until acceptable measurements are reached. Make sure that all feet take an equal load; unequal loading of the mounting bolts may distort the engine block. Engine alignment can be a time-consuming and frustrating business. When everything looks fine, tightening down the engine may throw out the alignment again. Patience is the order of the day.

Constant Velocity Joints (CVJs)

Marine-grade constant velocity joints are a refinement of the axle joints used in front-wheel drive cars. Two universal joints, with a short length of shaft between, are bolted between the coupling halves on a propeller shaft, permitting a considerable degree of engine misalignment (up to 1/2″ according to Aquadrive, the principal manufacturer; see Figure 9-20). Since the shaft is a sliding fit in one of the universal joints, a thrust bearing is also incorporated into the unit, and this must be fastened to a solid transverse bulkhead capable of absorbing full forward and reverse propeller loads. Without this, reverse propeller thrust would pull the unit apart.

CVJs are an unnecessary expense in most installations, but may prove to be the solution where irresolvable alignment problems are encountered. The transverse bulkhead needs careful design to avoid generating hard spots that could lead to hull damage. CVJs require no maintenance—the various bearings are packed in grease and sealed in rubber boots. However, since the principal components are steel, keep a

Figure 9-18. *Eliminating shaft droop. Attach a spring scale to the overhead; fix a line to the shaft and the scale and tension it until the scale reads one half the weight of the coupling.*

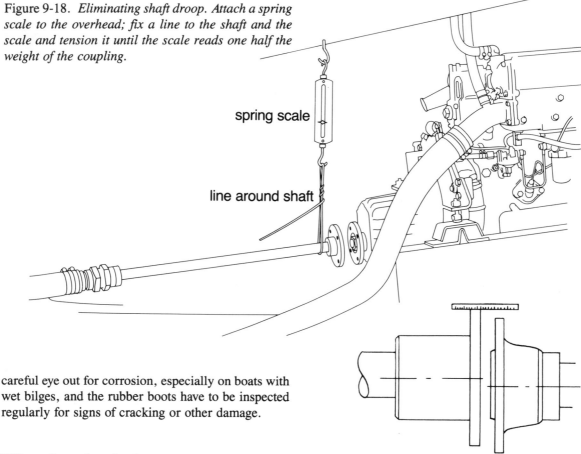

careful eye out for corrosion, especially on boats with wet bilges, and the rubber boots have to be inspected regularly for signs of cracking or other damage.

Vibration Analysis

The following is excerpted, with thanks, from the Caterpillar Tractor Company's *Marine Engines Application and Installation Guide*, published in October 1982, page 39.

The causes of linear vibrations can usually be identified by determining if:

1. The vibration amplitudes *increase with the speed*. If so, they are probably caused by centrifugal forces' bending components of the drive shafts. Check for unbalance and misalignment.
2. The vibrations *occur within a narrow speed range*. This normally occurs in equipment attached to the machinery—pipes, air cleaners, etc. When vibrations show a maximum amplitude or peak out at a narrow speed range, the

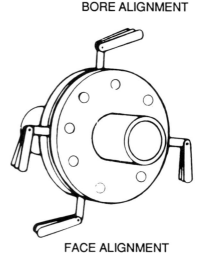

BORE ALIGNMENT

FACE ALIGNMENT

Figure 9-19. *Bore and face alignment. (Courtesy Caterpillar Tractor Co.)*

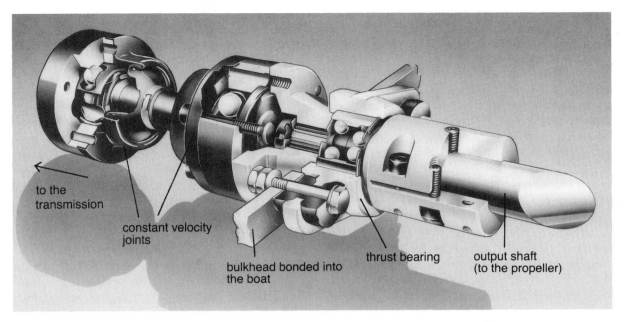

to the transmission

constant velocity joints

bulkhead bonded into the boat

thrust bearing

output shaft (to the propeller)

Figure 9-20. *Constant velocity joints, such as this Aquadrive unit, compensate for shaft misalignment.*

vibrating component is in *resonance*. These vibrations can be modified by changing the natural frequency of the part by stiffening or softening its mounting.

3. The vibrations *increase as load is applied*. This is caused by torque reaction and can be corrected by mounting the engine or driven equipment more securely or by stiffening the base or foundation. Defective or worn couplings can also cause this problem.

SECTION FIVE: AUXILIARY SYSTEMS

Ventilation

A diesel engine requires a large volume of clean *and* cool air (see Chapter 2). As air temperatures rise, the weight of air per cubic foot falls, and the engine pulls in correspondingly less oxygen at each cycle, causing a loss in efficiency and power.

Figure 9-21 gives an approximate idea of the decrease in the weight of air as the temperature rises. It is not uncommon for an engine room in the tropics to be as hot as 120°F (49°C), with turbocharging inlet–air temperatures considerably higher. Figure 9-22 shows the decrease in the engine's rated output as inlet air temperatures rise (as determined by the Diesel Engine Manufacturers Association—the rating starts at 90°F [32°C]). In excessively hot engine rooms, it will be necessary to duct air from outside directly to the air-inlet manifold. If such ducting is installed, its opening must be situated in a way that prevents water from entering it. Ducting also must be as far from the exhaust as possible so that the engine does not suck in spent gases.

It is also important, particularly on auxiliary sailboat engines housed in an insulated box or in a sealed engine room, to ensure an adequate flow of air to the engine. Otherwise, at higher loadings the engine is likely to be partially strangled, resulting in a loss of power, overheating, and probably black smoke from the exhaust.

Fuel Tanks and Filters

The density of fuel also decreases with rising temperatures. Fuel temperatures above 90 °F (32 °C—the rating point for American-made diesels) will result in reduced engine power. The higher a fuel tank is situated in an engine room, the warmer it will get. Keep tanks down low. In any event, the top of the tank should not be above the level of the injectors because occasionally fuel may leak down through the injectors into the combustion chambers. Low tanks will help to provide stability to the boat; centrally placed tanks will not affect athwartships trim.

If tanks are placed below an engine, all fuel connections can be made through the top of the tank. If a fuel line ruptures, this will prevent the entire contents of the tank from draining or siphoning into the bilges. In fact, there should be no openings, not even a drain, below the level of the tank top. If the tank needs to be emptied, pump it out from above.

The fuel fill line is best extended inside the tank since this reduces foaming when filling. The fuel suction line should extend no lower than 2 inches from the bottom of the tank in order to avoid sucking up contaminants. The distance from the bottom of a fuel tank to the fuel lift (feed) pump should be measured and compared to the lifting capacity of the pump. An auxiliary pump may be needed in certain exceptional cases.

All marine diesels without exception must have a primary filter installed in the fuel line between the tank and the engine-mounted lift (feed) pump. This filter must have a water separation capability and preferably a see-through bowl so that the operator can see at a glance if the fuel is contaminated. Baldwin/ Dahl and Racor manufacture an excellent line of primary filters to suit any engine. Powerboats should have dual filters mounted on a valved manifold that allows either filter to be changed without shutting down the engine.

The injector leak-off (return) lines from the engine should go directly to the tank. All too often they are led to the secondary (engine-mounted) fuel filter, and this can allow air to enter the fuel supply. Fuel lines must be of a properly approved fire retardant material and be adequately fastened against vibration. Avoid high spots where pockets of air can gather.

Diesel fuel tanks can be made of fire-retardant fiberglass, epoxy or glass-coated wood, plain black steel (not galvanized, at least on the inside, because sulfur traces in the fuel will dissolve the galvanizing, which can clog injectors), aluminum, or stainless steel. Aluminum and stainless steel are subject to cor-

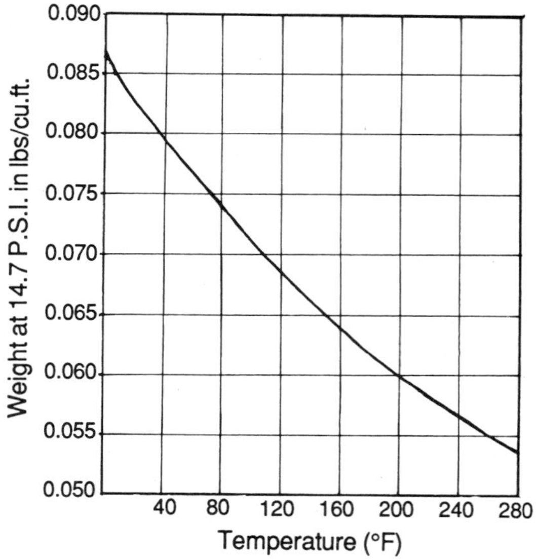

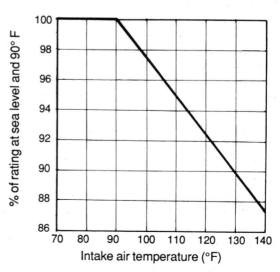

Figure 9-21. *Effect of temperature on weight of air.*

Figure 9-22. *Effect of air temperature on the performance of a diesel engine.*

rosion when they are in damp areas of poor air circulation, so you gain little benefit from the extra expense over a steel tank. Steel tanks *externally* hot-sprayed with zinc, primed with the appropriate paints, and sealed with a two-part epoxy paint, give excellent service.

Tanks need to be baffled at regular intervals in order to prevent the fuel from sloshing around; no compartment should contain more than 25 gallons. Access hatches for cleaning are needed into every compartment. Since contamination of the fuel supply is *the number one cause of problems with diesel engines*, have a small sump built into the base of the tank and provide some way to drain it or pump it out. Figure 9-23 illustrates a proper fuel-tank installation.

On wooden boats, you must ensure adequate air circulation between the tank and the hull against which it is mounted, or rot is likely to set into the woodwork (steel and aluminum hulls may also be susceptible to corrosion). It is often recommended that tanks be removable to permit proper hull inspections, but this is sometimes a very tall order to fill. At the least, the tank should be accessible enough to allow a close inspection of the woodwork around it.

Cooling

Almost all modern marine diesel engines have an enclosed freshwater cooling system, utilizing a heat exchanger or a keel cooler. (Keel coolers must be manufactured from marine-grade materials, such as those built by the Walter Machine Co. of Jersey City, N.J.) The combination of salt water, heat and the diverse metals found in some heat exchangers is a potent one for corrosion. *All heat exchangers should be protected with sacrificial zinc anodes.*

A raw-water seacock must be below the waterline at all angles of heel. It needs an adequate strainer (preferably see-through). The seacock must be made of marine-grade materials—bronze or one of the newer plastics, such as Marelon. It should be fastened by its skin fitting (the fitting on the outside of the hull that screws into the body of the seacock) and *with bolts through independent mounting flanges* (see Figure 9-24). The bolts prevent the seacock from falling

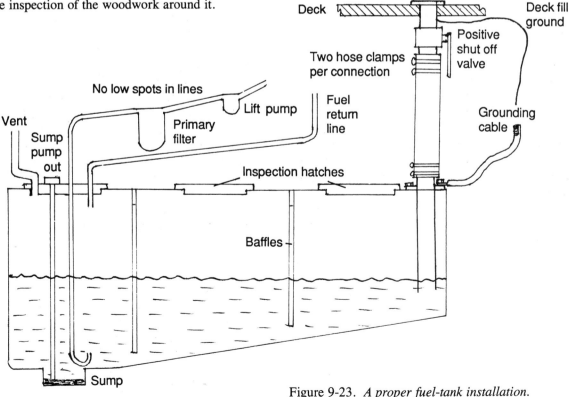

Figure 9-23. *A proper fuel-tank installation.*

Figure 9-24. *Proper through-hull and seacock installation, wood and fiberglass hulls (top). Proper installation in metal hulls (bottom).*

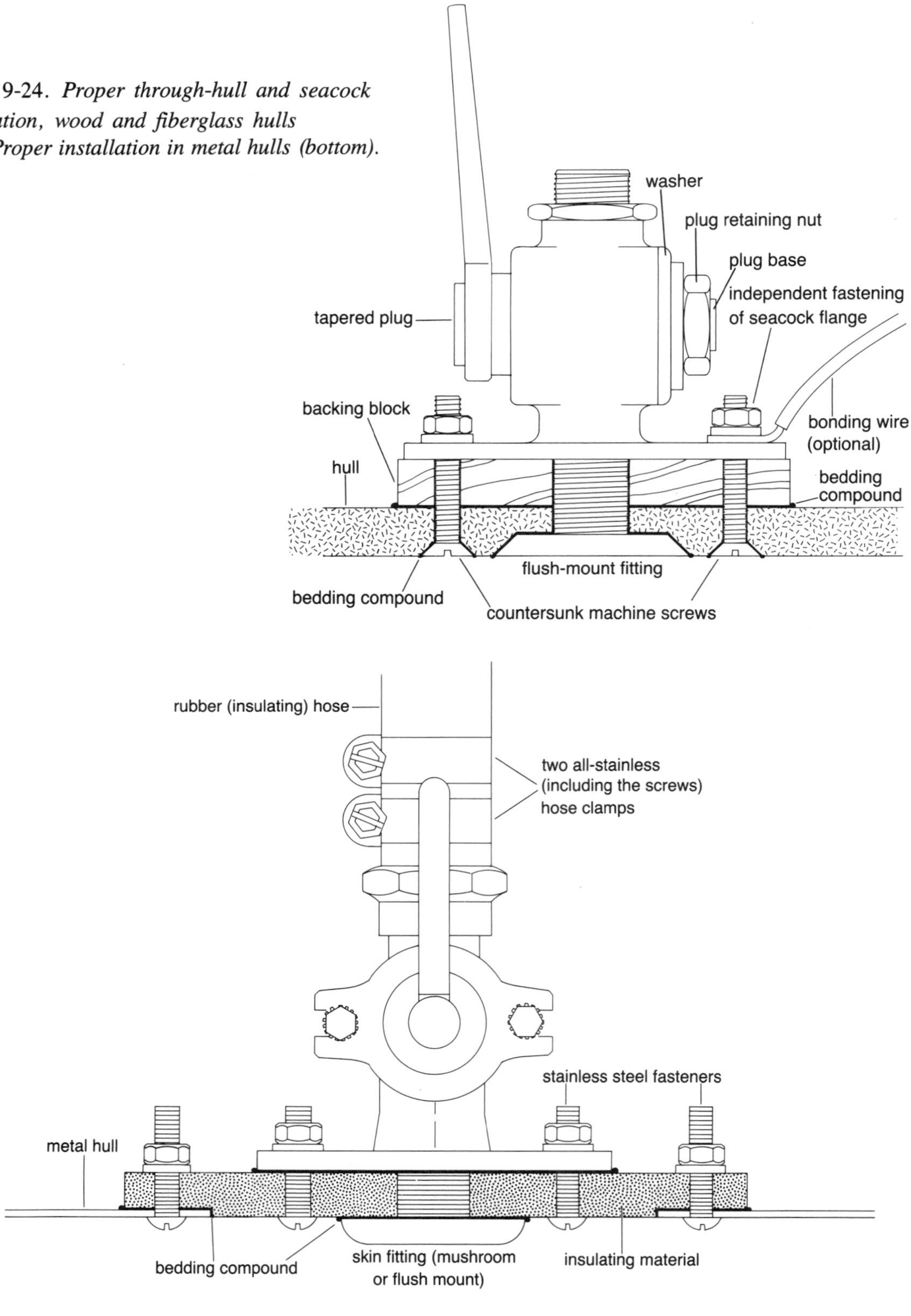

washer

plug retaining nut

plug base

independent fastening of seacock flange

tapered plug

bonding wire (optional)

backing block

hull

bedding compound

bedding compound

flush-mount fitting

countersunk machine screws

rubber (insulating) hose

two all-stainless (including the screws) hose clamps

stainless steel fasteners

metal hull

bedding compound

skin fitting (mushroom or flush mount)

insulating material

out of the hull if the skin fitting fails. Any kind of a gate valve is a very poor choice for a seacock, because these valves are prone to jamming and failure. A conventional seacock or a ball valve will give more reliable service.

When running raw-water hoses, avoid undrainable low spots where freezing might cause damage. All connections below the waterline should be double-clamped with stainless-steel hose clamps that have stainless screws.

Exhaust

With the exception of some 2-cycle diesels and certain special applications, most marine diesel engine exhausts are water cooled. This keeps the exhaust pipe cool and goes a long way towards reducing exhaust noise. Incorrect installations, however, have frequently caused water to siphon back into engines, doing expensive damage and even sinking boats, so it is necessary to look at this subject in some detail.

The idea is to spray water into the exhaust gases as they exit the engine. The water then collects in the base of a muffler (silencer) until it blocks the exit pipe. At this point the exhaust pressure rises until it is high enough to drive the water out of the exhaust pipe.

The water injection line must be at least 4 inches below the level of the engine exhaust manifold outlet, and angled down and away from the manifold, in order to reduce the risk of water splashing back onto the exhaust valves and rusting them (see Figure 9-25). Even so, there is a tendency for the valve nearest the exhaust outlet to rust—probably from steam coming back up the exhaust pipe after the engine is shut down.

The exhaust must continue to slope down from the water-injection point into the silencer, which should be at least 12″ below the manifold outlet. The silencer must be mounted *as close to the centerline of the boat as possible* in order to reduce the risk of water surging back into the engine when the boat heels to extreme angles.

Any silencer set this low in a boat is almost always well below the waterline. When the engine is shut down, water flowing into the exhaust pipe or through the water-injection line will steadily siphon into the silencer then back up the exhaust pipe into the mani-

fold. If any exhaust valves are open, the water will fill this cylinder and dribble down past piston rings into the crankcase. To prevent this, water-cooled exhausts must be fitted with the following safeguards:

1. *The raw-water injection line must be carried well above sea level (to the underside of the deck) and fitted with a siphon break.* The usual approach is to fit an anti-siphon valve—a simple device that admits air into the system and prevents siphoning action from developing. In salt water, however, these valves sometimes plug up and fail. An alternative approach is to fit a small tee at the top of the water-injection line and to vent it somewhere well above the waterline.

2. *The exhaust pipe too should be looped well above the waterline*, and vented clear of the water on the centerline of the boat so that it is not submerged on either tack. However, the higher an exhaust is looped, the greater the back pressure created in the exhaust because the gases must lift the water up the pipe. Back pressure impairs performance. The vertical lift from the muffler to the top of the pipe should not exceed 40″ on a naturally aspirated engine, and 20″ on a turbocharged engine. This latter dimension is hard to achieve (see below for an alternative).

3. *The exhaust should have a readily accessible cut-off valve* so that it can be closed in heavy following swells when the engine is not in use. If this is not done, waves driving up the back of the boat can steadily fill the exhaust. Even the repeated "water hammer" from water skiers in an otherwise calm anchorage can steadily drive water up some exhaust pipes. Many powerboats can also drive water up their exhausts by backing down hard.

Even with these precautions, if an engine is cranked repeatedly without firing, the raw-water pump will steadily fill a water-lift silencer while there will be no exhaust gases to drive this water out. With continued cranking the water can back up into the engine. A water-lift silencer needs a drain plug to deal with such a situation and for draining in the winter.

Sometimes there is insufficient space below the engine to get the silencer low enough to satisfy the requirements given above, or in order to keep the sea-

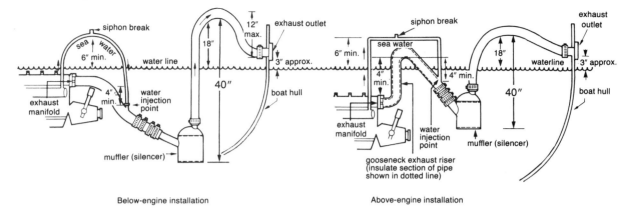

Figure 9-25. *Water lift muffler installations. Note: It would be preferable to fit a shut-off valve in the exhaust line, especially on sailboats, so that following seas can be prevented from driving up the exhaust pipe when the engine isn't running. (Courtesy Allcraft Corp.)*

water out of the exhaust, the exhaust must be looped up to a point at which it creates excessive back pressure. In this case, a modified wet silencer can be installed as illustrated, with the silencer set above the engine and the water injected directly into the silencer (see Figure 9-26). Because the exhaust gases do not have to lift the injected water, this system results in less back pressure (better for turbocharged engines and Detroit Diesel 2-cycles), but it will not silence the exhaust as effectively as the water-lift type. *The exhaust pipe leading into the silencer will be extremely hot and must be adequately insulated.*

Some form of insulated metal exhaust pipe is required for the dry sections of an exhaust. Beyond the water-injection point, good quality wire-reinforced steam hose is the best material. The hose needs to be securely fastened at regular intervals. Water-lift mufflers made of fire-retardant fiberglass do not suffer from the corrosion problems experienced by their metal counterparts.

Auxiliary Equipment

An increasing amount of extra equipment is driven off the engine these days, including alternators, bilge pumps, wash-down pumps, hydraulic pumps, refrigeration compressors, water-makers, and AC generators. The normal method of driving this equipment is

to fasten a small auxiliary *stub* shaft to the forward end of the crankshaft, install a couple of pulleys on it, and belt-drive the equipment from these pulleys. The following are caveats to keep in mind:

- Equipment mounted in this fashion exerts a lateral pull on the crankshaft. There is a strict limit to how much side-loading an engine can tolerate without damaging crankshaft oil seals and bearings. Check the manufacturer's specifications to see that this is not exceeded. (Note: placing loads on opposite sides of the engine partially cancels side-loading forces.)

- If the engine is flexibly mounted (most are) but auxiliary equipment is mounted *off the engine* (i.e., fastened to the hull or a bulkhead), this equipment can flex the engine on its feet, pulling it out of alignment with the propeller shaft. This flexing will also alter the drive-belt tension on the auxiliary equipment, which may cause problems.

- The driving pulleys on the engine stub shaft must be in alignment with the driven pulleys on the auxiliary equipment. A straight rod or dowel held in the groove of one pulley should drop cleanly into the groove on the other pulley (see Figure 9-27 for examples of misalignment).

- The auxiliary loads may interfere with the engine's cranking speed enough to make starting difficult in cold weather.

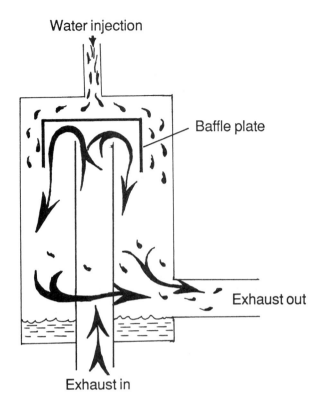

Water injection

Baffle plate

Exhaust out

Exhaust in

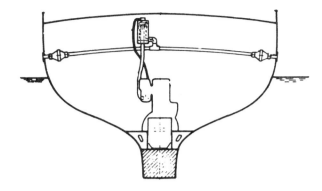

Figure 9-26. *A modified wet silencer (left). North Sea exhaust with alternative wet exhaust (right). (Courtesy Sabb Motor S.A.)*

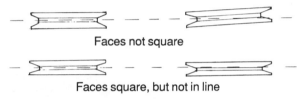

Faces not square

Faces square, but not in line

Figure 9-27. *Pulley alignment.*

Some Electrical Considerations

Diesel engines have compression ratios well above those of gasoline engines (two to three times higher). They are therefore harder to turn over. In spite of this increased resistance to cranking, turning over briskly is essential for successful starting, especially in cold weather. These two things taken together mean that *diesel engines require powerful batteries for effective starting*. You'd be wise to install a battery at least as powerful as the engine manufacturer recommends; additional capacity will be a bonus.

If an engine-cranking battery is also used to power the ship's electrics, it will steadily be discharged when the engine (and alternator) are shut down, then later recharged when the engine runs. This is called cycling. Any battery designed for engine cranking that is repeatedly cycled will soon fail. A deep-cycle battery will be needed.

Batteries are generally rated in amp-hours, a measurement of their ability to deliver a relatively low current over a 20-hour period (10 hours in the UK). Engine-cranking needs a very large current for a short period of time, and a rating known as cold-cranking amps (CCA) is used to measure this capability. Due to features of internal construction, a deep-cycle battery with the same amp-hour rating as a cranking battery will deliver fewer cold cranking amps: *you will need to buy a deep-cycle battery of larger overall capacity to give you the same cranking capability*. The key here is to check the cold-cranking-amps rating, rather than the amp-hour rating.

An engine-driven alternator is normally the primary charging device on a boat, once the boat is away from the dockside. It is not uncommon to see anchored-out boats running their engines two hours or more a day to recharge batteries. Diesels do not like this kind of prolonged low load running at near idling speeds.

If you plan to go cruising, you will save handsomely on fuel and repair bills in the long run if you buy a high-output alternator and a purpose-built marine voltage regulator with your new engine. The

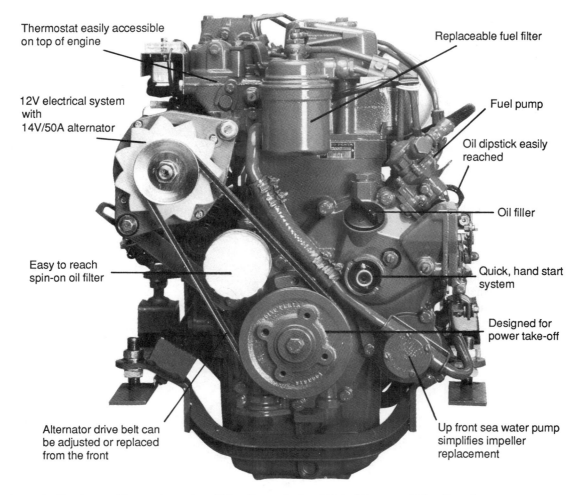

Thermostat easily accessible on top of engine

12V electrical system with 14V/50A alternator

Easy to reach spin-on oil filter

Alternator drive belt can be adjusted or replaced from the front

Replaceable fuel filter

Fuel pump

Oil dipstick easily reached

Oil filler

Quick, hand start system

Designed for power take-off

Up front sea water pump simplifies impeller replacement

Figure 9-28. *Accessible service points-Volvo Penta Series 2000. (Courtesy Volvo Penta)*

alternator will impose a heavier demand on the engine, while at the same time it keeps the low-load running hours to a minimum. If you can load up the engine some more when at anchor (e.g., with a refrigeration compressor), then so much the better. Even better still would be to take care of your charging needs altogether with a wind generator, solar panels, or both. (See my *Boatowner's Mechanical and Electrical Manual* for a detailed discussion of all these subjects.)

Safety Shut-downs

Dirty fuel is the number-one cause of diesel engine problems; engine overheating and low oil pressure are other common difficulties. Proper maintenance procedures will eliminate most causes of these problems, while a variety of engine alarms and shut-downs can be fitted to minimize the damage if you fail to catch the causes.

You can buy fuel filters, which incorporate two

electric probes. If water accumulates in the filter and rises to the level of the probes, it closes a circuit and activates an alarm and/or shut-down device. Engine-temperature and oil-pressure gauges can be fitted with a terminal, which the needle in the gauge contacts, setting off an alarm, shut-down device, or both when the temperature reaches a certain level, or the oil pressure drops to a certain point. Alternatively, you can fit dual sensing devices (incorporating both a gauge circuit and an alarm circuit) to the engine block, to monitor both the temperature and the oil pressure and to trigger an alarm if either approach dangerous levels.

An alarm circuit is simply wired to a bell, whereas a shut-down circuit generally incorporates a solenoid valve on the fuel injection pump. When the circuit is activated, the solenoid closes off the fuel to stop the engine (usually activating an alarm bell at the same time). I prefer having only an alarm circuit—there may be times when the safety of the ship and boat demand that the captain ignore a warning signal and risk the destruction of the engine in order to escape danger.

Serviceability

As I have repeatedly pointed out, two things are absolutely critical to the long life of a diesel engine: clean fuel and clean oil. If you keep the fuel uncontaminated and properly filtered, and you change the oil and filter at the prescribed intervals (generally every 100-150 running hours), most diesels will run for years without giving trouble.

An engine installation needs to be designed to simplify these items of basic maintenance (see Figure 9-28). The fuel and oil filters must be readily accessible, and the engine oil easy to change. Changing the oil in most boats requires an oil-change pump because the drain plug in the engine pan is generally hard to get at, and even if you can remove it, there is no room to slip a suitable container under the pan to catch the oil (see page 35).

The raw-water pump must be accessible, if only to be able to loosen its cover in order to drain it when you winterize the boat. And what about any grease points? The stuffing box is another item to think about—so often you have to slide in head first through a cockpit locker and hang upside down in the bilges to get at it. What if the engine develops other problems? I once worked on a boat in which the entire engine had to be pulled off its mounts in order to change the starter motor. If the engine needs a major overhaul, will you be able to remove the cylinder head? The time to think about these things is now, not when you are faced with a breakdown out in the boonies.

Appendix A

Tools

A basic set of mechanics tools is required for working on engines—wrenches, a socket set (preferably ½″ drive), screwdrivers, hacksaw, crescent wrench, Vise Grips, etc. The tool kit should also include a copy of the appropriate manufacturer's shop manual. This appendix covers one or two more specialized items.

1. **Oil squirt can.** Preferably with a flexible tip, for putting oil into the air inlet manifold.

2. **Grease gun.** Again preferably with a flexible hose.

3. **Feeler (thickness) gauges.** From 0.001″ to 0.025″ (or the metric equivalent if one's clearances are specified in millimeters).

4. **Oil filter clamp** for spin-on-type filters. Such filters are extremely difficult to get on and off without this special purpose tool. More than one size may be required if the fuel filters are a different size.

5. **Grinding paste.** This is sold in three grades: coarse, medium, and fine. There is very little call for coarse, and the medium and fine can very often be bought in one container.

6. **Suction cup** and **handle** for lapping in valves.

7. **Torque wrench.** An indispensable tool for any serious mechanical work, and probably the most expensive special item, although there are some perfectly serviceable and relatively inexpensive ones on the market for occasional use. The wrench will fit ½″ drive sockets or other sizes with suitable adaptors.

8. **Ball peen hammer.** Most people have carpenters hammers with a jaw for pulling nails, but in mechanical work, a ball peen hammer is far more useful. Hammers are specified by the weight of the head—an 8-oz. hammer is a good all-around size.

9. **Needlenose pliers.** Handy for all kinds of tasks—side-cutting needlenose pliers also have a wire-cutting jaw, and are preferred.

10. **Scrapers.** For cleaning up old gaskets.

11. **Mallet** or **soft-faced hammer.** A surprisingly valuable tool, especially if one has to knock an aluminum or cast-iron casting that might be cracked by a steel hammer.

12. **Aligning punches.** Invaluable from time to time, especially the long ones (8″ to 10″). These punches are tapered—a light one (with a tip around ⅛″) and a heavier one (around ¼″) will do nicely.

13. **Injector bar.** An injector bar is about 15″ long, tapered to a point at one end, and with a heel on the other end. It is a very useful tool for prying or levering.

14. **Allen wrenches.** Almost certainly required at some point. Keep an assortment on hand.

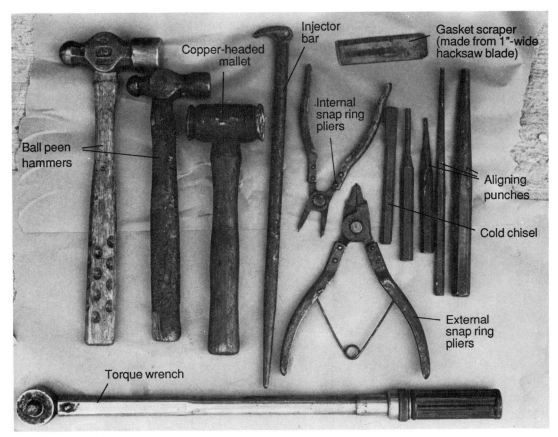

Figure A-1. *Useful tools.*

15. **Hydrometer.** Needed for testing batteries. It is best to get one of the inexpensive plastic ones, since the regular glass ones, though more accurate, break sooner or later.

16. **Snap-ring pliers.** We are now getting into the realm of very specialized equipment for the serious mechanic. Snap rings can almost always be gotten out with needlenose pliers, regular pliers, or the judicious use of screwdrivers.

17. **Valve-spring clamp.**

18. **Piston-ring expander.** This can be dispensed with as indicated in the text.

19. **Piston-ring clamp.**

20. **Gear puller.** These come in all shapes and sizes. There are a number of gears and pulleys in any engine that just cannot be removed, unless you use some kind of a puller, without risking damage to engine castings or other parts. It is frequently possible to improvise, as indicated in the text.

What should never be done is to put levers behind a gear to try and force it off—the effort generally ends in failure, and frequently in damage to some casting, the gear, the pulley, or the shaft.

21. **Injector nozzle cleaning set.** Includes a brass-bristle brush and the appropriate nozzle hole prickers for one's injectors. Might be a worthwhile investment for the long-distance cruiser. CAV and others sell these sets with appropriate tools for their own injectors.

A good tool kit represents a considerable expense but will last a lifetime if cared for. In general, *it is not worth buying cheap tools:* sooner or later they break, but long before this point they drive you crazy by slipping and bending. There are quite enough problems in engine work without creating any unnecessary ones.

Appendix B

Spares

The extent of your spare parts inventory will obviously depend on your cruising plans. The following is a fairly comprehensive list with an ocean cruising sailboat in mind. If your plans are less ambitious it can be scaled down appropriately.

1. Oil filters.
2. Fuel filters—quite a number in case a dirty batch of fuel is taken on board and repeated filter changes are needed.
3. An air filter—one only. This should rarely need changing in the marine environment.
4. A raw-water pump overhaul kit, or at the very least, a diaphragm or impeller (depending on the type of pump).
5. A lift-pump overhaul kit, or at the very least, a diaphragm.
6. A fuel injection pump diaphragm if one is fitted.
7. A complete set of belts (alternator, plus auxiliary equipment).
8. An alternator.
9. A starter motor solenoid and Bendix unit.
10. A cylinder head overhaul gasket set.
11. An inlet and exhaust valve.
12. Two sets of valve springs and keepers (exhaust and inlet, if they differ).
13. A set of piston rings.
14. A set of connecting rod bearing shells.
15. One or two injectors, or at the very least matched sets of replacement needle valves and seats and nozzles.
16. A complete set of high-pressure injection lines (from the fuel injection pump to the injectors).
17. A complete engine gasket set.
18. A gearbox oil seal.
19. An O ring kit with an assortment of O rings may one day be worth its weight in gold.
20. A roll of high-temperature gasket paper.
21. A roll of cork-type gasket paper.
22. A tube of gasket compound.
23. Packing for the stern tube stuffing box (Teflon or flax, but not graphite, since this can cause electrolysis).
24. Assorted hose clamps.
25. Flexible fuel line.
26. Hoses.
27. Oil and grease.
28. Penetrating oil.
29. Exhaust elbow.

Appendix C

Useful Tables

This is a rather mixed bag of information that may come in handy at some time or another.

1 HP = 33,000 foot-pounds per minute (550 foot-pounds per second).

$$\text{Torque (in foot-pounds)} = \frac{\text{BHP} \times 5,250}{\text{r.p.m.}}$$

$$\text{BHP} = \frac{\text{Torque} \times \text{r.p.m.}}{5,250}$$

1 Btu = 778 foot-pounds.

1 HP = 2,544 Btus.

1 KW = 1.34 h.p.

1 KW = 3,413 Btus.

100 cubic inches = 1.64 liters.

1 gallon (U.S.) of pure water weighs 8.34 lbs.

1 cubic foot of pure water weighs 62.4 lbs.

1 imperial gallon (U.K.) = 1.2 gallons (U.S.).

Circumference of a circle = $2 \pi R$ or πD, where π = 3.14.

Area of a circle = πR^2.

Volume of a cylinder = $\pi R^2 \times$ length.

$$°C = \frac{(°F - 32) \times 5}{9}$$

$$°F = \frac{(°C \times 9)}{5} - 32$$

1 short ton (U.S.) = 2,000 lbs.

1 long ton (U.K.) = 2,240 lbs.

Inches	Millimeters	Inches	Millimeters	Inches	Millimeters
0.001	0.0254	0.010	0.2540	0.019	0.4826
0.002	0.0508	0.011	0.2794	0.020	0.5080
0.003	0.0762	0.012	0.3048	0.021	0.5334
0.004	0.1016	0.013	0.3302	0.022	0.5588
0.005	0.1270	0.014	0.3556	0.023	0.5842
0.006	0.1524	0.015	0.3810	0.024	0.6096
0.007	0.1778	0.016	0.4064	0.025	0.6350
0.008	0.2032	0.017	0.4318		
0.009	0.2286	0.018	0.4572		

Figure C-1. *Inches to millimeters conversion table.*

Torque Conversion, Pound Feet/Newton Metres

Pound-Feet (lb.-ft.)	Newton Metres (Nm)	Newton Metres (Nm)	Pound-Feet (lb.-ft.)
1	1.356	1	0.7376
2	2.7	2	1.5
3	4.0	3	2.2
4	5.4	4	3.0
5	6.8	5	3.7
6	8.1	6	4.4
7	9.5	7	5.2
8	10.8	8	5.9
9	12.2	9	6.6
10	13.6	10	7.4
15	20.3	15	11.1
20	27.1	20	14.8
25	33.9	25	18.4
30	40.7	30	22.1
35	47.5	35	25.8
40	54.2	40	29.5
45	61.0	50	36.9
50	67.8	60	44.3
55	74.6	70	51.6
60	81.4	80	59.0
65	88.1	90	66.4
70	94.9	100	73.8
75	101.7	110	81.1
80	108.5	120	88.5
90	122.0	130	95.9
100	135.6	140	103.3
110	149.1	150	110.6
120	162.7	160	118.0
130	176.3	170	125.4
140	189.8	180	132.8
150	203.4	190	140.1
160	216.9	200	147.5
170	230.5	225	166.0
180	244.0	250	184.4

Figure C-2. *Metric conversion table.*

Fraction, decimal, and metric equivalents

Fractions	Decimal In.	Metric mm.	Fractions	Decimal In.	Metric mm.
1/64	.015625	.397	33/64	.515625	13.097
1/32	.03125	.794	17/32	.53125	13.494
3/64	.046875	1.191	35/64	.546875	13.891
1/16	.0625	1.588	9/16	.5625	14.288
5/64	.078125	1.984	37/64	.578125	14.684
3/32	.09375	2.381	19/32	.59375	15.081
7/64	.109375	2.778	39/64	.609375	15.478
1/8	.125	3.175	5/8	.625	15.875
9/64	.140625	3.572	41/64	.640625	16.272
5/32	.15625	3.969	21/32	.65625	16.669
11/64	.171875	4.366	43/64	.671875	17.066
3/16	.1875	4.763	11/16	.6875	17.463
13/64	.203125	5.159	45/64	.703125	17.859
7/32	.21875	5.556	23/32	.71875	18.256
15/64	.234375	5.953	47/64	.734375	18.653
1/4	.250	6.35	3/4	.750	19.05
17/64	.265625	6.747	49/64	.765625	19.447
9/32	.28125	7.144	25/32	.78125	19.844
19/64	.296875	7.54	51/64	.796875	20.241
5/16	.3125	7.938	13/16	.8125	20.638
21/64	.328125	8.334	53/64	.828125	21.034
11/32	.34375	8.731	27/32	.84375	21.431
23/64	.359375	9.128	55/64	.859375	21.828
3/8	.375	9.525	7/8	.875	22.225
25/64	.390625	9.922	57/64	.890625	22.622
13/32	.40625	10.319	29/32	.90625	23.019
27/64	.421875	10.716	59/64	.921875	23.416
7/16	.4375	11.113	15/16	.9375	23.813
29/64	.453125	11.509	61/64	.953125	24.209
15/32	.46875	11.906	31/32	.96875	24.606
31/64	.484375	12.303	63/64	.984375	25.003
1/2	.500	12.7	1	1.00	25.4

Appendix D

Freeing Frozen Fasteners

Problems with frozen fasteners are inevitable on boats. One or more of the following techniques may free things up.

Lubrication

- Clean everything with a wire brush (preferably one with brass bristles), douse liberally with penetrating oil, and wait. Find something else to do for an hour or two, overnight if possible, before having another go. Be patient.
- Clevis pins: After lubricating and waiting, grip the large end of the pin with Vise-Grips (Mole wrench) and turn the pin in its socket to free it. If the pin is the type with a cotter pin (also known as a cotter key or split pin) in both ends, remove one of the cotter pins, grip the clevis pin, and turn. Since the Vise-Grips will probably mar the surface of the pin, it should be knocked out from the other end.

Shock Treatment

An impact wrench is a handy tool to have around. These take a variety of end fittings (screwdriver bits; sockets) to match different fasteners. The wrench is hit hard with a hammer and hopefully jars the fastener loose. If an impact wrench is not available or does not work, other forms of shock must be applied with an acute sense of the breaking point of the fastener and adjacent engine castings, etc. Unfortunately this is generally only acquired after a lifetime of breaking things! Depending on the problem, shock treatment may take different forms:

- A bolt stuck in an engine block: Put a sizable punch squarely on the head of the bolt and give it a good knock into the block. Now try undoing it.
- A pulley on a tapered shaft, a propeller, or an outboard motor flywheel: Back out the retaining nut *until its face is flush with the end of the shaft* (this is important to avoid damage to the threads on the nut or shaft). Put pressure behind the pulley, propeller, or flywheel as if trying to pull it off, and hit the end of the retaining nut or shaft smartly. The shock will frequently break things loose without the need for a specialized puller.
- A large nut with limited room around it, or one on a shaft that wants to turn (for example, a crankshaft pulley nut): Put a short-handled wrench on the nut, hold the wrench to prevent it from jumping off, and hit it hard.
- If all else fails, use a cold chisel to cut a slot in the side of the offending nut or the head of the bolt, place a punch in the slot at a tangential angle to the nut or bolt, and hit it smartly.

Leverage

- Screws: With a square-bladed screwdriver, put a crescent (adjustable) wrench on the blade, bear down hard on the screw, and turn the screwdriver with the wrench. If the screwdriver has a round blade, clamp a pair of Vise-Grips to the base of the handle and do the same thing.

- Nuts and bolts: If using wrenches with one box end and one open end, put the box end of the appropriate wrench on the fastener and hook the box end of the next size up into the free open end of the wrench to double the length of the handle and thus the leverage.
- Cheater pipe: Slip a length of pipe over the handle of the wrench to increase its leverage.

Heat

Heat expands metal, but for this treatment to be effective, frozen fasteners must frequently be raised to cherry-red temperatures. These temperatures will upset tempering in hardened steel, while uneven heating of surrounding castings may cause them to crack. Heat must be applied with circumspection.

Heat applied to a frozen nut will expand it outward, and it can then be broken loose. But equally, heat applied to the bolt will expand it within the nut, generating all kinds of pressure that helps to break the grip of rust, etc. When the fixture cools it will frequently come apart quite easily.

Broken Fasteners

- Rounded-off heads: Sometimes there is not enough head left on a fastener to grip with Vise-Grips or pipe (Stillson) wrenches, but there is enough to accept a slot made by a hacksaw. A screwdriver can then be inserted and turned as above.
- If a head breaks off it is often possible to remove whatever it was holding, thus exposing part of the shaft of the fastener, which can be lubricated, gripped with Vise-Grips, and backed out.
- Drilling out: It is very important to drill down the center of a broken fastener. Use a center punch and take some time putting an accurate "dimple" at this point before attempting to drill. Next use a small drill to make a pilot hole to the desired depth. If Ezy-Outs or "screw extractors" (hardened, tapered steel screws with reversed threads, available from

tool supply houses) are on hand, drill the correctly sized hole for the appropriate Ezy-Out and try extracting the stud. Otherwise drill out the stud *up to the inside of its threads but no farther,* or irreparable damage will be done to the threads in the casting. The remaining bits of fastener thread in the casting can be picked out with judicious use of a small screwdriver or some pointed instrument. If a tap is available to clean up the threads, so much the better.

- Pipe fittings: If a hacksaw blade can be gotten inside the relevant fittings (which can often be done using duct tape to make a handle on the blade), cut a slit in the fitting along its length, and then place a punch on the outside alongside the cut, hit it, and collapse it inward. Do the same on the other side of the cut. The fitting should now come out easily.

Miscellaneous

- Stainless steel: Stainless-to-stainless fasteners (for example, many turnbuckles) have a bad habit of "galling" when being done up or undone, especially if there is any dirt in the threads to cause friction. Galling (otherwise known as "cold welding") is a process in which molecules on the surface of one part of the fastener transfer to the other part. Everything seizes up for good. Galled stainless fastenings cannot be salvaged—they almost always end up shearing off. When doing up or undoing a stainless fastener, if any sudden or unusual friction develops stop immediately, let it cool off, lubricate thoroughly, work the fastener backward and forward to spread the lubrication around, go back the other way, clean the threads, and start again.
- Aluminum: Aluminum oxidizes to form a dense white powder. Aluminum oxide is more voluminous than the original aluminum and so generates a lot of pressure around any fasteners passing through aluminum fixtures—sometimes enough pressure to shear off the heads of fasteners. Once oxidation around a stainless or bronze fastener has reached a certain point it is virtually impossible to remove the fastener without breaking it.
- Damaged threads: If all else fails, and a fastener has to be drilled out, the threads in the casting may

be damaged. There are two options.

1. To drill and tap for the next larger fastener.
2. To install a Heli-Coil insert, a Heli-Coil is a new thread. An oversized hole is drilled and tapped with a special tap, and the Heli-Coil insert (the new thread) is screwed into the hole with a special tool. You end up with the original sized hole and threads. Any good machine shop will have the relevant tools and inserts.

Glossary

Aftercooler. Also called an intercooler. A heat exchanger fitted between a turbocharger and an engine air-inlet manifold in order to cool the incoming air.

Alignment. The bringing together of two coupling halves in near-perfect horizontal and vertical agreement.

Ambient. The surrounding temperature, pressure, or both.

Annealing. A process of softening metals.

Atmospheric pressure. The pressure of air at the surface of the earth, conventionally taken to be 14.7 psi.

Atomization. The process of breaking up diesel fuel into minute particles as it is sprayed into an engine cylinder.

Babbitt. A soft white metal alloy frequently used to line replaceable shell-type engine bearings.

Back pressure. A build-up of pressure in an exhaust system.

BHP (Brake Horsepower). The actual power output of an engine at the flywheel.

Bleeding. The process of purging air from a fuel system.

Blow-by. The escape of gases past piston rings or closed valves.

Bottom Dead Center (BDC). A term used to describe the position of a crankshaft when the #1 piston is at the very bottom of its stroke.

Btu (British thermal unit). The unit used to measure quantities of heat.

Butterfly valve. A hinged flap connected to a throttle that is used to close off the air inlet manifold on gasoline engines and some diesel engines.

Cams. Elliptical protrusions on a camshaft.

Camshaft. A shaft with cams, used to operate the valve mechanism on an engine.

Cavitation. The process by which a propeller sucks down air and loses contact with the water in which it is turning.

Circlips. See snap rings.

Collets. See keepers.

Combustion chamber. The space left in a cylinder (and cylinder head) when a piston is at the top of its stroke.

Common rail. A type of fuel injection system in which fuel circulates to all the injectors all of the time. Each injector contains its own injection pump with this system.

Compression ratio. The volume of a compression chamber with the piston at the top of its stroke as a proportion of the total volume of the cylinder when the piston is at the bottom of its stroke.

Connecting rod. The rod connecting a piston to a crankshaft.

Connecting rod bearing. The bearing at the crankshaft end of a connecting rod.

Connecting rod cap. The housing that bolts to the end of a connecting rod, holding it to a crankshaft.

Crank. An offset section of a crankshaft to which a connecting rod is attached.

Cranking speed. The speed at which a starter motor turns over an engine.

Crankshaft. The main rotating member in the base of an engine, transmitting power to the flywheel and power train.

Cutlass bearing. A rubber-sleeved bearing in the stern of a boat that supports the propeller shaft.

Cylinder block. The housing on an engine that contains the cylinders.

Cylinder head. A casting containing the valves and injector that bolts to the top of a cylinder block and seals off the cylinders.

Cylinder liner. A machined sleeve that is pressed into a cylinder block and in which a piston moves up and down.

Decarbonizing. The process of removing carbon from the inside surfaces of an engine and of refurbishing the valves and pistons.

Decompression levers. Levers that hold the exhaust valves open so that no compression pressure is built up, making it easy to turn the engine over.

Dial indicator. A sensitive measuring instrument used in alignment work.

Displacement. The total swept volume of an engine's cylinders expressed in cubic inches or liters.

Distributor pump. A type of fuel injection pump using one central pumping element with a rotating distributor head that sends the fuel to each cylinder in turn.

Dribble. Drops of unatomized fuel entering a cylinder through faulty injection.

Feeler (thickness) gauges. Thin strips of metal machined to precise thicknesses and used for measuring small gaps.

Flyweight. A small pivoted weight used in mechanical governors.

Friction Horsepower. The proportion of the power generated by an engine consumed in the operation of the engine itself (from friction, and from driving water and injection pumps, camshafts, etc.).

Fuel injection pump. A device for metering precise quantities of fuel at precise times and raising them up to injection pressures.

Garboard. The side plank closest to the keel on a wooden boat.

Gasket. A piece of material placed between two engine parts to seal them against leaks. Gaskets are normally fiber but sometimes metal, cork, or rubber.

Glow plugs. Heating elements installed in precombustion chambers to assist in cold starting.

Governor. A device for maintaining an engine at a constant speed, regardless of changes in load.

Head gasket. The gasket between a cylinder head and a cylinder block.

Heat exchanger. A vessel containing a number of small tubes through which the engine cooling water is passed, while raw water is circulated around the outside of the tubes to carry off the engine heat.

Header tank. A small tank set above an engine on heat-exchanger-cooled systems. The header tank serves as an expansion chamber, coolant reservoir, and pressure regulator (via a pressure cap).

Hole-type nozzle. An injector nozzle with one or more very fine holes—generally used in direct (open) combustion chambers.

Horsepower. A unit of power used in rating engines.

Hunting. Cyclical changes in speed around a set point, usually caused by governor malfunction.

Hydrometer. A tool for measuring specific gravity.

Inches of mercury. A scale for measuring small pressure changes, particularly those below atmospheric pressure (i.e., vacuums).

Indicated Horsepower. The actual power developed by an engine before taking into account internal power losses.

Injection timing. The relationship of the beginning point of injection to the rotation of the crankshaft.

Injector. A device for atomizing diesel fuel and spraying it into a cylinder.

Injector nozzle. That part of an injector containing the needle valve and its seat.

Injector nut. The nut that holds a fuel line to an injector.

Intercooler. See aftercooler.

In-line pump. A series of jerk pumps in a common housing operated by a common camshaft.

Jerk pump. A type of fuel injection pump that uses a separate pumping element for each cylinder.

Keepers. Small, dished, metal pieces that hold a valve spring assembly on a valve stem.

Lapping. A process of grinding two parts together to make an exact fit.

Lift pump. A low pressure pump that feeds diesel fuel from a tank to an injection pump.

Line contact. The machining of two mating surfaces at different angles so that they make contact only at one point.

Main bearing. A bearing within which a crankshaft rotates, and which supports the crankshaft within an engine block.

Manifold. A pipe assembly attached to an engine block that conducts air into the engine or exhaust gases out of it.

Micrometer. A tool for making precision measurements.

Naturally aspirated. Refers to an engine that draws in air solely by the action of its pistons, without the help of a supercharger or turbocharger.

Needle valve. The valve in an injector nozzle.

Nozzle body. The housing at the end of an injector that contains the needle valve.

Nozzle opening pressure. The pressure required to lift an injector needle valve off its seat so that injection can take place.

Pintle nozzle. An injector nozzle with one central hole—generally used in engines with precombustion chambers.

Piston. A pumping device used to generate pressure in a cylinder.

Piston crown. The top of a piston.

Piston pin (wrist pin). A pin connecting a piston to its connecting rod, allowing the piston to oscillate around the rod.

Piston rings. Spring-tensioned rings set in grooves in the circumference of a piston that push out against the walls of its cylinder to make a gastight seal.

Piston-ring clamp. A tool for holding piston rings tightly in their grooves to enable the piston to be slid into its cylinder.

Piston-ring groove. The slot in the circumference of a piston into which a piston ring fits.

Ports. Holes in the wall of a cylinder that allow gases in and out.

Pounds per square inch absolute (absolute pressure). Actual pressure measurements with no allowance for atmospheric pressure.

Pounds per square inch gauge (gauge pressure—psi). Pressure measurements taken with the gauge set to zero at atmospheric pressure (14.7 psi absolute).

Power train. Those components used to turn an engine's power into a propulsive force.

Pumping losses. Energy losses arising from friction in the inlet and exhaust passages of an engine.

Push rod. A metal rod used to transfer the motion of a camshaft to a rocker arm.

Pyrometer. A gauge used to measure exhaust temperatures.

Raw water. The water in which a boat is floating.

Rocker arm. A pivoted arm that operates a valve.

Rocker cover. See valve cover.

Rod-end bearing. The bearing at the crankshaft end of a connecting rod.

Scavenging. The process of replacing the spent gases of combustion with fresh air in a 2-cycle diesel.

Seizure. The process by which excessive friction brings an engine to a halt.

Sensible heat. The temperature of a body as measured by a thermometer.

Shim. A specially cut piece of shim stock used as a spacer in specific applications.

Shim stock. Very thin, accurately machined pieces of metal.

Shaft Horsepower (SHP). The actual power output of an engine and power train measured at the propeller shaft.

Slip. The difference between the theoretical movement of a propeller through the water and its actual movement.

Snap rings. Spring-tensioned rings that fit into a groove on the inside of a hollow shaft or around the outside of a shaft.

Solenoid. An electrically operated valve or switch.

Specific gravity. The density of a liquid as compared to that of water.

Speeder spring. The spring in a governor that counterbalances the centrifugal force of the flyweights.

Stroke. The movement of a piston from the bottom to the top of its cylinder.

Stuffing box. A device for making a water-tight seal around a propeller shaft at the point where it exits a boat.

Supercharger. A mechanically driven blower used to compress the inlet air.

Swept volume. The volume of a cylinder displaced by a piston in one complete stroke (i.e., from the bottom to the top of its cylinder).

Thermostat. A heat-sensitive device used to control the flow of coolant through an engine.

Thrust bearing. A bearing designed to take a load along the length of a shaft (as opposed to perpendicular to it).

Timing. The relationship of valve and fuel pump operation to the rotation of the crankshaft and to each other.

Top Dead Center (TDC). A term used to describe the position of a crankshaft when the #1 piston is at the very top of its stroke.

Torque. A twisting force applied to a shaft.

Torque wrench. A special wrench that measures the force applied to a nut or bolt.

Turbocharger. A blower driven by an engine's exhaust gas that is used to compress the inlet air.

Vacuum. Pressure below atmospheric pressure.

Valves. Devices for allowing gases in and out of a cylinder at precise moments.

Valve clearance. The gap between a valve stem and its rocker arm when the valve is fully closed.

Valve cover. The housing on an engine that is bolted over the valve mechanism.

Valve guide. A replaceable sleeve in which the valve stem fits and slides up and down.

Valve keepers. See keepers.

Valve overlap. The period of time in which an exhaust valve and inlet valve are both open.

Valve seat. The area in a cylinder head on which a valve sits in order to seal that head.

Valve spring. The spring used to hold a valve in the closed position when it is not actuated by its rocker arm.

Valve-spring clamp. A special tool to assist in removing valves from their cylinder heads.

Viscosity. The resistance to flow of a liquid (its thickness).

Volatility. The tendency of a liquid to evaporate (vaporize).

Volumetric efficiency. The efficiency with which a 4-cycle diesel engine replaces the spent gases of combustion with fresh air.

Wrist pin. See piston pin.

Yoke. The hinged and forked lever arm that couples a governor flyweight to its drive shaft.

Index